Nucleic Acids and Molecular Biology

20

Series Editor
H. J. Gross

H. Ulrich Göringer (Ed.)

RNA Editing

 Springer

Professor Dr. H. Ulrich Göringer
Institute of Microbiology and Genetics
Technical University Darmstadt
Schnittspahnstr. 10
64287 Darmstadt
Germany
goringer@hrzpub.tu-darmstadt.de

ISSN 0933-1891
ISBN-13: 978-3-540-73786-5   e-ISBN-13: 978-3-540-73787-2

Library of Congress Control Number: 2007932404

Springer is a part of Springer Science+Business Media

springer.com

Printed on acid-free paper   5 4 3 2 1 0

Preface

"Editing a book on editing" has a certain "tongue in cheek" connotation to it, and I must admit that I was not aware what the job truly entailed when the project started. However, over the many months that it took to complete the book, it was interesting to realize that indeed "printing-type editing" and "biological editing" have many matching characteristics. Therefore, the quote by the well-known film editor Verna Fields (1918–1982), *"I wish the word "editing" had never been invented, "editing" implies correcting, and it's not"*, can function as a motto for the book: the various biological phenomena that are controlled by the different editing reactions are by far more complex than a simple correction process, and it is astounding how multifaceted the field has become.

All chapters of the book focus, as a general "editorial" subtext, on the correlation between RNA structure and function, and on complexity. This involves different length scales: from complex molecular machineries, to the interplay of complex biochemical pathways, and to evolutionary processes. The chapters are organized in a sequence beginning with a status quo account of RNA editing reactions from the focused perspective of RNA structure. This is followed by a chapter on the structure/function correlation in tRNA editing, and a chapter on RNA editing by adenosine deaminases that act on RNA (ADARs). A to I editing is perhaps today's best example for the wide spectrum of biological functions that are impacted by RNA editing. Especially the recently discovered interaction of the ADAR editing pathway with the microRNA/RNAi machinery is an excellent illustration of this. The next chapter summarizes the remarkable progress in the insertion/deletion-type editing reaction in *Physarum polycephalum*, and the following two chapters are dedicated to editing processes in plant mitochondria and chloroplasts. Three chapters deal with the U insertion/U deletion-type RNA editing reaction in kinetoplastid parasites. This incorporates a description of the molecular editing machinery, a chapter on accessory factors of the reaction pathway, and a detailed discussion of its biological function including the recently discovered alternative editing phenomenon. A final chapter is dedicated to evolutionary aspects of RNA editing.

I hope that the book will prove useful not only to those working directly in this domain, but also to advanced students and researchers, including those in sister disciplines, who wish to "re-edit" their knowledge on RNA editing.

Finally, I would like to thank all contributing authors for their work and efforts, and the series editor H.J. Gross for his help and advice. A special thank you goes to Ursula Gramm for her patient and calm editorial work, and to Monique T. Delafontaine for her superb copy-editing: a perfect example of editing as a "*conditio sine qua non*".

Darmstadt, June 2007   Uli Göringer

Contents

Contributors

Juan D. Alfonzo
Department of Microbiology and The Ohio State Biochemistry Program, The Ohio State University, 484 West 12th Avenue, Columbus, Ohio 43210, USA, e-mail: alfonzo.1@osu.edu

Cordula Böhm
Genetics, Darmstadt University of Technology, Schnittspahnstr. 10, 64287 Darmstadt, Germany

Michael Brecht
Genetics, Darmstadt University of Technology, Schnittspahnstr. 10, 64287 Darmstadt, Germany

Axel Brennicke
Molekulare Botanik, Universität Ulm, 89069 Ulm, Germany

Jason Carnes
Seattle Biomedical Research Institute, 307 Westlake Ave N, Suite 500, Seattle, WA 98109, USA, and Department of Pathobiology, University of Washington, Seattle, WA 98195, USA

H. Ulrich Göringer
Genetics, Darmstadt University of Technology, Schnittspahnstr. 10, 64287 Darmstadt, Germany, e-mail: goringer@hrzpub.tu-darmstadt.de

Jonatha M. Gott
Center for RNA Molecular Biology, 10900 Euclid Avenue, Case Western Reserve University, Cleveland, OH 44106, USA, e-mail: jmg13@case.edu

Stephen Hajduk
Department of Biochemistry and Molecular Biology, University of Georgia, 120 Green Street, Athens, GA 30602-7229, USA, e-mail: shajduk@bmb.uga.edu

Matthias Homann
Genetics, Darmstadt University of Technology, Schnittspahnstr. 10, 64287 Darmstadt, Germany, e-mail: mhomann@hrzpub.tu-darmstadt.de

Michael F. Jantsch
Department of Chromosome Biology, Max F. Perutz Laboratories, University of Vienna, Dr. Bohr Gasse 1, 1030 Vienna, Austria, e-mail: Michael.Jantsch@univie.ac.at

Elisabeth Kruse
Genetics, Darmstadt University of Technology, Schnittspahnstr. 10, 64287 Darmstadt, Germany

Julia Neuwirt
Molekulare Botanik, Universität Ulm, 89069 Ulm, Germany

Torsten Ochsenreiter
Department of Biochemistry and Molecular Biology, University of Georgia, 120 Green Street, Athens, GA 30602-7229, USA

Marie Öhman
Department of Molecular Biology & Functional Genomics, Stockholm University, 106 91 Stockholm, Sweden

Amy C. Rhee
Center for RNA Molecular Biology, 10900 Euclid Avenue, Case Western Reserve University, Cleveland, OH 44106, USA

Dave Speijer
Academic Medical Center (AMC), Department of Medical Biochemistry, University of Amsterdam, Meibergdreef 15, 1105 AZ Amsterdam, The Netherlands, e-mail: d.speijer@amc.uva.nl

Kenneth Stuart
Seattle Biomedical Research Institute, 307 Westlake Ave N, Suite 500, Seattle, WA 98109, USA, and Department of Pathobiology, University of Washington, Seattle, WA 98195, USA, e-mail: kstuart@u.washington.edu

Masahiro Sugiura
Graduate School of Natural Sciences, Nagoya City University, Nagoya 467-8501, Japan, e-mail: sugiura@nsc.nagoya-cu.ac.jp

Mizuki Takenaka
Molekulare Botanik, Universität Ulm, 89069 Ulm, Germany, e-mail: mizuki.takenaka@uni-ulm.de

Daniil Verbitskiy
Molekulare Botanik, Universität Ulm, 89069 Ulm, Germany

Dré van der Merwe
Molekulare Botanik, Universität Ulm, 89069 Ulm, Germany

Anja Zehrmann
Molekulare Botanik, Universität Ulm, 89069 Ulm, Germany

Editing Reactions from the Perspective of RNA Structure

Matthias Homann

> *"If something is in me which can be called religious then it is the unbounded admiration for the structure of the world so far as our science can reveal it."*
>
> Albert Einstein

Abstract RNA editing belongs to the large group of processing reactions that are required to convert primary RNA transcripts into mature and functional transcripts. The main determinants of specificity rest in the three-dimensional structures of RNA and protein molecules that act in concert to coordinate and regulate the posttranscriptional steps in gene expression. Many high-resolution structures of RNA–protein complexes, including the ribosome, have become available during the last decade and have offered detailed views of the intracellular RNA world. The focus of this review is to highlight the contributions of RNA structure to the specificity and efficiency of RNA editing. Editing occurs by a variety of mechanisms, but the fidelity of the reactions critically depends on the specific sequences and structures of the RNA molecules involved and on their recognition by trans-acting factors, including proteins and RNA. Hence, the editing machineries, also termed "editosomes", make

Genetics, Darmstadt University of Technology, Schnittspahnstr. 10, 64287 Darmstadt, Germany;
mhomann@hrzpub.tu-darmstadt.de

use of RNA–RNA, RNA–protein and protein–protein interactions to achieve specificity and efficiency. High-resolution structures of protein components of various editosomes exist, but reports of RNA structures and RNA–protein complexes are still limited. Progress can be expected in the near future from more efficient purification and crystallization techniques developed in other fields of RNA processing, like RNA interference, splicing and catalysis. Although each structure reveals only a static view of a multistep reaction, they will eventually lead to a better understanding of the dynamic molecular machines involved in RNA editing.

1   Introduction

Eukaryotic gene expression involves extensive processing of primary RNA transcripts to generate mature and functional RNAs. A wide variety of RNA binding proteins act in concert with complex three-dimensional structures of RNA molecules to coordinate and regulate posttranscriptional steps in gene expression. High-resolution structures of RNA–protein complexes as well as of single components involved in RNA modification (reviewed in Reichow et al. 2007), RNA splicing (Stark and Lührmann 2006), RNA catalysis (RNase P: Evans et al. 2006; group I introns: Vicens and Cech 2006; small ribozymes: Nelson and Uhlenbeck 2006), RNA transport (Müller et al. 2007) and RNA interference mechanisms (Ma et al. 2004, 2005; Song et al. 2004; Parker et al. 2005; MacRae et al. 2006) have become available during the last decade. Above all, the crystal structures of the ribosome and its subunits have offered unprecedented views of the intracellular RNA world (for reviews, see Moore and Steitz 2003; Noller 2005) and allow predictions as to the dynamics and the molecular interactions that underlie RNA-based gene expression pathways.

The focus of this review is to highlight the contributions of RNA structure to the specificity and efficiency of RNA editing. RNA editing is a widespread phenomenon in which the sequence information of certain transcripts is altered by enzymatic modification or the addition/deletion of nucleotides. Editing occurs by a variety of mechanisms, but a minimal set of components can be defined that are essential for accurate editing. First, the RNA to be edited usually contains sequence and/or structural information needed to specify the editing site ("cis-acting signals"). Second, trans-acting factors, which can be proteins as well as RNA, have to interact with the target RNA at or near the editing site. Third, enzymatic activities have to be directed to the specific site and catalyze the editing reaction(s). All reaction steps are catalyzed within RNP complexes (reviewed in Moore 2005), and the accuracy of editing critically depends on the specific RNA sequences and structures and on their recognition by the protein factors. These RNP complexes have in some cases been termed "editosome", in analogy with the two major RNA-based molecular machines, the ribosome and the spliceosome.

1.1 Principles of RNA Structure and RNA–Protein Interactions

Owing to the wealth of high-resolution structures of the ribosome (e.g. Ban et al 2000; Wimberly et al. 2000; Hansen et al. 2002; Korostelev et al. 2006), many principles of RNA folding have been uncovered (reviewed in Noller 2005). Local RNA motifs such as U-turns, T-loops, K-turns, A-minor interactions, A platforms and tetraloops have been classified (Leontis and Westhoff 2003; Lescoute et al. 2005; Leontis et al. 2006) and are likely to contribute to RNA editing processes as well. However, only one of the RNAs/RNP complexes involved in RNA editing has been crystallized so far (Schumacher et al. 2006). In addition, the solution structure of the RNA substrate for a specific C-to-U deamination editing event is available (see Sect. 2.4; Maris et al. 2005a).

RNA–protein interactions are governed by a distinct set of protein motifs that are involved in RNA binding (Chen and Varani 2005; see Table 1). The most common domains are the RNA recognition motif (RRM, also called RNA binding domain, RBD, or ribonucleoprotein domain, RNP), the K-homology (KH) domain, and the double-stranded RNA binding domain, dsRBD. In humans, the RRM alone is present in about 500 proteins, or 2% of the human genome, often found in multiple copies or in combination with other domains (reviewed in Maris et al. 2005b). Another very common domain is the zinc-finger motif that accounts for about 3% of the human genome, although only few zinc-fingers of the CCCH and CCHC type are known to be involved in RNA recognition (Lu et al. 2003; Hudson et al. 2004). Table 1 lists several other motifs that are present in specific enzymes acting on RNA (e.g. RNase III) or that function in specific processes such as RNA interference (e.g. PAZ and PIWI domains). Some particularly striking examples of RNA recognition are provided by proteins using helical repeat motifs, such as the Pumilio proteins (Wang et al. 2002), TRAP (Antson et al. 1999) and Sm/Lsm proteins (Kambach et al. 1999). In addition, small domains with highly basic stretches rich in Arg and Lys have been found to be involved in RNA recognition, e.g. in the cases of HIV-1 Tat (Ye et al. 1995) and Rev (Battiste et al. 1996) proteins. These domains, however, are not strictly pre-organized motifs, since they will fold into their active structure only upon binding to RNA (Frankel and Smith 1998). High-resolution structures of the most common motifs have been determined, either in their isolated form or in complex with RNA. These structures generally reveal a highly conserved folding of the respective motif, with little conservation at the amino acid sequence level. Accordingly, these motifs function as platforms capable of high-affinity recognition of practically any RNA sequence. This principle is best illustrated by the large number of proteins using the RRM

Table 1 Protein motifs involved in RNA recognition

Motif	Appearance	Reference
RRM, RBD, RNP	Sxl	Handa et al. (1999)
	PAB	Deo et al. (1999)
	U1A	Allain et al. (1997)
	U2B/U2A	Price et al. (1998)
	hnRNPA1	Ding et al. (1999)
	Nucleolin	Johansson et al. (2004)
KH	Nova	Lewis et al. (2000)
DsRBD	Staufen	Ramos et al. (2000)
	RNase III	Kharrat et al. (1995)
PAZ	Dicer	MacRae et al. (2006)
	Ago	Ma et al. (2004)
PIWI	Ago	Ma et al. (2005)
Helical repeat	Pumilio	Wang et al. (2002)
	TRAP	Antson et al. (1999)
	Sm/Lsm proteins	Kambach et al. (1999)
Arg-rich	Tat	Puglisi et al. (1995)
	Rev	Battiste et al. (1996)
Zn2+ finger	TFIIIA	Lu et al. (2003)
	TIS11d	Hudson et al. (2004)

domain for RNA recognition (Maris et al. 2005b), which may even be used for protein–protein interactions (Kielkopf et al. 2004).

Most of the RNA binding motifs listed in Table 1 have been identified in proteins known to be involved in diverse RNA editing mechanisms (e.g. Stuart et al. 2005). However, given the large variability of RNA recognition mechanisms, the identification of any of these motifs in putative editing components is of little predictive value to define protein function.

One of the most elegant ways for proteins to recognize RNA sequences of any kind is the use of RNA as co-factors: short RNA molecules with partial complementarity to the respective target RNAs. This principle can be found in the cases of U1, U2 and U6 RNAs that are part of the snRNPs and support splice site definition as well as catalysis (Valadkhan 2005). Other examples are the modification guide snoRNAs that act on pre-rRNAs (Reichow et al. 2007), and siRNAs as well as microRNAs that target corresponding mRNAs for degradation or translational arrest (Valencia-Sanchez et al. 2006). The most complex of all editing mechanisms, the kinetoplastid mitochondrial editing, makes use of such RNA co-factors (see Sect. 2.6 and chapters by Carnes and Stuart, Göringer et al., and Ochsenreiter and Hajduk, all in this volume). Indeed, the term "guide RNA" has been coined for the short 60–75 nt RNA molecules that specify editing sites during kinetoplastid editing (Blum et al. 1990).

2 Editing of mRNA Sequences

2.1 Paramyxovirus

Members of the Paramyxoviridae group are non-segmented negative-strand RNA viruses (NNV) that contain five to ten linked genes on a single RNA genome of 15 to 16 kb. Virus-encoded RNA-dependent RNA polymerases (RdRP) terminate transcription of individual mRNAs by stuttering on short runs of uridylate residues at the 3′ end of each gene, which effectively leads to mRNA polyadenylation (Lamb and Kolakofsky 1996). This stuttering of the RdRP is also responsible for editing of the viral phosphoprotein (P-) mRNA by the insertion of a distinct (and virus-specific) number of G-nucleotides at a single defined site of the mRNA (Vidal et al. 1990). For Sendai virus, the co-transcriptional insertion of a single G-nucleotide occurs along the "slippery" template sequence 3′-UUUUUUC$\underline{C}^{1052}$C-5′ (U_6C_3) once the G-nucleotide opposite C^{1052} is added to the growing mRNA. The RdRP pauses, allowing the AAAAAAGG message to slip back one nucleotide such that the six template Us pair with five As and the first G-nucleotide. C^{1052} is copied again as the RdRP resumes elongation, resulting in the net insertion of one G-nucleotide (Hausmann et al. 1999a).

Two features of the RNA sequence around and upstream (relative to the growing mRNA) of the editing site contribute to the pausing and stuttering of the viral RdRP (Hausmann et al. 1999b). First, the "slippery" polypyrimidine template sequence allows realignment of template and message without interrupting the duplex bound to the active centre of the RdRP. The thermodynamic barrier to this realignment will thus be very low (Kolakofsky and Hausmann 1998). Second, the number of Gs inserted is determined genetically for each paramyxovirus by a six-nucleotide sequence immediately upstream of the U_6C_3/A_6G_3 duplex. This region supposedly lines the exit channel of the RdRP and, therefore, is likely to be engaged in base-specific RNA–protein interactions with the RdRP (Hausmann et al. 1999a). Although detailed structural information of this interaction is not available, the conservation of the six-nucleotide sequence within the group of the Paramyxoviridae argues for a crucial function of this sequence in determining the target site and the number of G-insertions.

2.2 Physarum

RNA editing in *Physarum polycephalum* is a mitochondrial process that affects mRNA, rRNA and tRNA transcripts alike. At least two distinct editing mechanisms operate in the mitochondria. The first reaction is a C-to-U conversion at four positions within the transcript of the cytochrome C oxidase subunit 1, co1 (Gott et al. 1993). This type of editing resembles plant mitochondrial editing, since it usually affects the first or second codon position and restores conserved protein sequence (Gott 2001). The second mechanism is the insertion of more than 400 C-residues, less than 100 U-residues and several dinucleotide combinations (UU, AA, UA, CU, GU, GC) into nearly all mitochondrial RNA transcripts (Gott et al. 2005). These insertional editing events cause numerous frame shifts that are required to restore correct reading frames in most of the mitochondrial mRNAs. Recently, nucleotide deletion editing was identified in the transcript of *nad2* where three consecutive adenines were missing in the final mRNA (Gott et al. 2005). The insertion (and, potentially, also the deletion?) of nucleotides is clearly a co-transcriptional event that is either a feature of the mitochondrial RNA polymerase itself or is promoted by additional trans-acting factors that associate with the polymerase (Cheng et al. 2001). Mono- and dinucleotide insertions may be directed by a distinct set of specificity factors (Wang et al. 1999; Horton and Landweber 2000; Byrne and Gott 2004).

Editing site selection is the result of local features of the RNA sequence immediately preceding the editing site (Byrne and Gott 2002) and their interaction with the editing machinery, consisting of – at least – the mitochondrial RNA polymerase. Additional specificity factors such as trans-acting guide RNAs or proteins have not been identified as of yet but may be involved in modulating polymerase activity. Therefore, the detailed mechanisms of editing site recognition and subsequent nucleotide insertion still remain to be clarified. The RNA polymerases itself may play the dominant role, since it is more closely related to the single-chain bacteriophage polymerases (Masters et al. 1987; Cermiakan et al. 1996). These enzymes exhibit some editing-related activities, such as the addition of non-templated nucleotides within homopolymer tracts (Macdonald et al. 1993) or at the 3′ termini of transcripts, e.g. by the T7 RNA polymerase (Milligan et al. 1987).

Editing sites in mRNA transcripts show a number of statistical rules that point to a function of local RNA primary sequence in editing site selection (Gott et al. 2005). On average, insertion events occur every 25 nucleotides on a given mRNA, with two thirds located at the third codon position. Cis-acting RNA signals are restricted to the region less than 15 bases upstream of the editing site (Byrne and Gott 2002). However, statistical analysis of the base composition within this region revealed no significant sequence pattern (Gott et al. 2005). The only conserved bases are found at pos. −1 being uridine in 82% of the editing sites, and pos. −2 being a purine (62% A, 21% G), combining for a purine-U motif at −2/−1 in 69% of all cases. Co-variation analysis of pairs of bases within the region from −15 to +15 surrounding the editing sites revealed no correlations between pairs of bases. Thus, the formation of conserved Watson-Crick base pairs as part of RNA secondary structures in the vicinity of editing sites is unlikely. Since nucleotide insertions are co-transcriptional events, DNA template sequences surrounding the editing site may contribute to editing specificity. Potential cis-acting DNA template determinants were shown to be restricted to a region within 15 bp upstream and 15–20 bp of downstream DNA (Byrne and Gott 2002).

Since the purine/U bias at the −2/−1 position upstream of the editing site is the only cis-acting element common to all editing sites, additional specificity factors must be present at the site of transcription. Either a guiding mechanism involving guide RNAs (that should involve at least one guide per site with the potential to form at least 7–8 bp) can be postulated, or additional proteins act as specificity factors by interacting with the nascent RNA and/or by modulating polymerase activity and processivity. Identification of such RNA as well as protein co-factors will require the ability to isolate and purify editosome complexes from editing active mitochondrial fractions that could be subjected to mass spectroscopy. Recently, a putative mitochondrial RNA polymerase from *Physarum polycephalum* was identified (Miller et al. 2006) that may serve as a starting protein for the affinity purification of editosome complexes using the Tap-tagging technology (Rigaut et al. 1999).

2.3 RNA Editing in Plant Organelles

RNA editing in the mitochondria and chloroplasts of plants occurs by C-to-U deamination that affects a large number of mRNAs encoded by the organellar genomes. In a given plant species, about 20–40 Cs are deaminated in the mRNAs of chloroplasts (Maier et al. 1996; Tsudzuki et al. 2001), contrasting with the 400–1,000 editing sites in mRNAs within the mitochondria. In addition, very few (~0.5%) reverse U-to-C editing events have been identified. About 90% of all editing events affect the coding regions of mRNAs and, in most cases, the editing reaction is essential for the restoration of conserved codons/amino acids.

Given the large number of plant organellar editing events, it seems surprising that the specificity factors are still elusive. Several recent analyses in the mitochondria of pea (Takenaka et al. 2004), wheat (Choury et al. 2004; Choury and Araya 2006) and cauliflower (van der Merwe et al. 2006), and in the chloroplasts of tobacco (Chateigner-Boutin and Hanson 2002; Hayes et al. 2006) and *Arabidopsis* (Hegeman et al. 2005) have narrowed the search for the cis-acting RNA signals to the immediate vicinity of the editing site. The essential elements usually reside within 15–20 nts upstream and 1–5 nucleotides downstream of the editing site. However, these flanking regions do not share consensus elements at the primary or secondary structure level, suggesting that site-specific editing requires several hundreds of specific factors. In a recent study, the editing efficiencies at two sites, C77 and C259, within the wheat mitochondrial cox2 transcript were compared (Choury et al. 2004). In both cases, essential sequences were identified within a 22-nt region from −16 to +6 relative to the editing site (+1). This 22-nt sequence could be placed at different sites within the large RNA without loss of editing efficiency, suggesting context-independent editing. Replacement of the downstream region (+2 to +6) of C77 with the homologous region from the C259 site led to strongly reduced editing efficiency, whereas replacement of the upstream region (−16 to −1) completely abolished editing at C77. These results confirm that each editing site sequence is recognized by an individual set of specificity factors.

Some of these factors may recognize groups of related sequences (Chateigner-Boutin and Hanson 2002, 2003), and editing sites with identical core determinants (−15 to +2) exist that are likely to be recognized by the same trans-acting factor (Tillich et al. 2005). Recently, two protein candidates for site-specific trans-acting factors were identified by cross-linking technology (Miyamoto et al. 2002, 2004). A third protein factor was identified in *Arabidopsis* chloroplasts by a genetic screening method (Kotera et al. 2005). In this study, mutants were isolated that were defective in a unique RNA editing event that creates the initiation codon of the chloroplast NAD(P)H dehydrogenase subunit D (ndnD). The mutations were assigned to the CRR4 protein, a member of the plant combinatorial and modular protein family (PCMP; Aubourg et al. 2000) that is a subgroup of the penta-tricopeptide repeat (PPR) family (Small and Peeters 2000). The PPR family consists of about 450 genes in the *Arabidopsis* genome. Members of the PPR family are involved in RNA processing, stabilization and translation in chloroplasts and mitochondria of yeasts and higher plants (Kotera et al. 2005 and references therein). The PPR repeats have been implicated in RNA binding in the case of the chloroplast PPR protein HCF152 (Nakamura et al. 2003), and the number of PPR repeats was shown to determine the affinity and specificity for RNA. CRR4 itself contains 11 PPR repeats consisting of a degenerate 35-amino acid unit, and Kotera et al. (2005) suggest it functions as a specificity factor for the recognition of the first editing site in the ndhD transcript. The authors propose a model, similar to the apoB editing system (see Sect. 2.4), in which PPR proteins interact with individual editing sites to recruit the editing machinery containing the deaminase activity. Consistent with this model, 189 of the 452 PPR proteins in *Arabidopsis* were predicted to localize to the mitochondria, and 96 to the chloroplasts (Nakamura et al. 2003).

PPR proteins belong to the family of helical repeat proteins, and the modes of RNA recognition may be analogous to those of another helical repeat protein family, the PUF proteins. The five PUF proteins in yeast were recently shown to regulate groups of 100–200 target mRNAs carrying similar 3′ UTR recognition motifs of 9–11 nucleotides (Gerber et al. 2004). The crystal structure of the human PUF protein Pumilio1 in complex with its RNA target revealed that such RNA sequence motifs are bound as single-stranded unstructured RNA. Each base within the sequence motif is contacted by a single Pumilio homology domain (PHD, ~34 amino acids) via 4–5 amino acid side chains. Two of the side chains are required for base-specific hydrogen bonding to the Watson-Crick interface, while two hydrophobic side chains form a "sandwich" with the base stacking in between. Thus, the helical repeats of PPR may functionally resemble those of the PUF proteins and may serve as a platform for interaction with a variety of RNA molecules. The sequence specificity of RNA recognition should rely on distinct amino acid side chains exposed on this platform to contact the Watson-Crick interface of the RNA bases.

Following this model, the PPR proteins would direct an editing deaminase activity to the RNA target site that is bound in a single-stranded and unstructured form. The protein might even serve to pre-orient the editing site C towards deamination, in analogy with the function of the ACF protein during apoB-editing described in the following section.

2.4　Mammalian Editing: apoB

Editing of apolipoprotein B (apoB) mRNA was the first RNA editing event described in vertebrates (Chen et al. 1987; Powell et al. 1987). The editing reaction is catalyzed by the APOBEC1 enzyme (apoB editing catalytic subunit 1), the first identified mammalian cytidine deaminase (Teng et al. 1993). APOBEC1 is expressed in the small intestine, where it catalyzes the formation of a premature stop codon in the apoB mRNA by deamination of cytidine^{6666} to uridine. As a result, a truncated form of apoB with 48 kDa (apoB48) is expressed, which has a tissue-specific function in lipid metabolism in the small intestine that differs from that of its full-length counterpart apoB100 expressed in the liver. The editing event thus contributes to genetic plasticity by generating two protein isoforms from a single gene locus. APOBEC1 belongs to a family of 12 cytidine deaminases identified so far in the human genome (reviewed in Turelli and Trono 2005). Most of these are involved in the deamination of dC to dU at the level of DNA, with diverse cellular functions in antibody diversification, retroviral defence and retrotransposon silencing (reviewed in Holmes et al. 2007).

2.4.1　APOBEC-1 Target Site Structure and Dynamics

APOBEC-1 is the only member of the deaminase family for which an mRNA target could be identified. The target site of APOBEC1 lies within the 14,000-nt apoB mRNA of which a single cytidine residue at position 6666 is recognized and deaminated by the editosome machinery. The specificity of the reaction relies on sequence motifs surrounding the editing site that are highly conserved within vertebrate species. The minimal editing competent sequence is 26 nucleotides long and folds into a stem-loop secondary structure that has important functions in the mechanism of RNA recognition by the editing factors (Richardson et al. 1998). It contains three cis-acting elements in addition to the C^{6666} editing site (Fig. 1). The first element is an 11-nucleotide mooring sequence located downstream of the editing site. The mooring sequence is separated from the C^{6666} editing site by a "spacer element" with a variable length of 2–8 nucleotides. The third sequence is an A/U-rich efficiency element upstream of the editing site that modulates editing efficiency. The minimal editing competent complex that recognizes the RNA stem-loop consists of two proteins: APOBEC1 as the catalytic subunit, and APOBEC1 complementation factor, ACF (Mehta et al. 2000). The 64-kDa ACF contains three

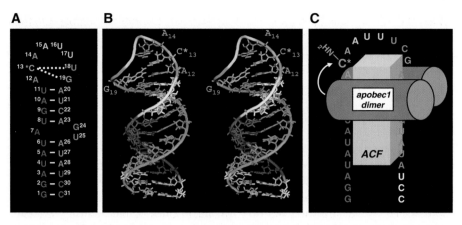

Fig. 1 RNA editing by site-specific deamination of C^{6666} in the apoB mRNA. **A** Secondary structure of a minimal 31-nt substrate of APOBEC1-dimer/ACF complex (Maris et al. 2005a). The editing site cytidine C^*13 is indicated (*cyan*; corresponds to C^{6666} of the apoB message). Sequence elements important for binding and site specificity are colour-coded (*green* mooring sequence, *red* efficiency element, *yellow* spacer element, *brown* remaining nts). **B** Stereo-view of the apoB stem-loop structure derived from NMR analysis (Maris et al. 2005a). One of the calculated structures is shown in which the target site C^* is sandwiched between A14 and A12 (both *orange*), and involved in hydrogen bonding to G19 (*cyan*). **C** ACF was shown to melt the apoB stem-loop in order to disrupt the interactions of the target site C^* and expose the NH_2 group for attack by an APOBEC1 dimer that is recruited to the target site by ACF (Maris et al. 2005a)

RNA recognition motifs (RRMs) and a putative double-stranded RNA binding domain (dsRBD). ACF binds to both APOBEC1 and the mooring sequence of the stem-loop RNA with high affinity (Kd = 8 nM, Mehta and Driscoll 2002), and is required to anchor APOBEC1 at the editing site. APOBEC1 itself was shown to interact with the sequence motif UUUN A/U U (Anant and Davidson 2000) 2 nts downstream of C6666 with much lower affinity. The enzyme has catalytic activity on a minimal apoB RNA substrate in the absence of auxiliary factors (Chester et al. 2004). The reaction has a temperature optimum of 45°C that was lowered to physiological temperatures in the presence of ACF. A model was suggested in which ACF promotes a conformational transition in the RNA substrate and stabilizes an editing competent conformation that forms spontaneously at higher temperatures (Chester et al. 2004).

The molecular basis of ACF function was addressed by NMR spectroscopy of the 31-nucleotide apoB-mRNA stem-loop containing the essential cis-acting elements (Maris et al. 2005a). The calculated structures confirmed a previously proposed model (Chester et al. 2004) and suggested the requirement for several conformational rearrangements of the RNA structure during the editing process. The authors proposed a model in which the rearrangements are induced by the sequential binding of ACF and APOBEC1.

The calculated structures for the unedited (C^{6666}) as well as the edited version (U^{6666}) of the apoB mRNA target reveal an extended stem-loop conformation in which the editing site nucleotide (C/U13) is located within a flexible octaloop (Richardson et al. 1998). The base is sandwiched between two adenines (A12 and A14) and involved in H-bond interactions with one of the two 3′-terminal loop-bases (G19 or U18; dotted lines in Fig. 1A). The remaining nucleotides of the octaloop (A15–G19) were disordered in the NMR structures and interpreted to be flexible (Maris et al. 2005a). Thus, the N4 position of C6666 is buried within the octaloop and a conformational change is required in order to expose the amino group for deamination.

The structural dynamics of the mooring sequence and its recognition by the ACF protein were proposed to contribute to this loop rearrangement (Maris et al. 2005a). In all known species, the mooring sequence is embedded in a structurally dynamic environment consisting of an irregular stem-loop structure with bulges and internal loops. In the NMR structure, the first two nucleotides of the mooring sequence (U18 and G19) are part of the octaloop and appear to be flexible. The remaining residues (A20 to A28), together with U4 to U11, form an irregular A-form helix interrupted by an internal loop composed of G24, U25 and A7. This internal loop has a flexible conformation that also destabilizes the surrounding AU base pairs and even the hydrogen bond interactions of the upper stem base pairs. Thus, most of the mooring sequence is very dynamic (U18, G19) to moderately dynamic (A20 to A26). This flexibility plays a critical role in RNA recognition, since a mooring sequence completely annealed to its antisense strand (Mehta and Driscoll 2002) is not recognized by ACF. The unpaired nucleotides U18, G19, G24 and U25 could serve as a primary recognition site for ACF. Thus, it seems that editing of apoB mRNA requires a flexible mooring sequence, a flexible spacer and no strong base pairing of the editing site C with surrounding base pairs.

In order to analyse the role of ACF in promoting the necessary conformational changes, RNA binding studies were performed with ACF34, the N-terminal part of ACF containing the three essential RRMs (Maris et al. 2005a). NMR studies indicated that ACF34 bound to U18 and U25 at 25 °C without denaturing the stem-loop. At 37–42 °C, however, the stem-loop was melted, indicating that ACF34 bound the RNA as a single strand. By melting the RNA stem-loop, ACF34 also disrupts the structure around C13 and renders its amino group accessible for deamination by APOBEC-1. In addition, interaction site mapping by UV cross-linking studies revealed that binding of ACF to the RNA restricted the access of APOBEC-1 to a single site (C13) for editing (Maris et al. 2005a). These experimental findings support the conclusions drawn from homology modelling of APOBEC-1 using the coordinates of the yeast cytidine deaminase CCD1 crystal structure as a template (Xie et al. 2004). The resulting APOBEC-1 dimer model revealed that the RNA substrate would only be accommodated in its single-stranded form, supporting an essential function of ACF in disrupting the RNA hairpin secondary structure.

The sequence of events outlined above can be viewed as a hallmark example for the structural and conformational changes that occur during a given editing event: formation of a well-defined and, at the same time, flexible RNA structure; site-specific binding of a protein factor; creation of a nucleation site for melting of the stem-loop structure; recruitment of the enzymatic editing activity and restricting its

access to a single defined site; catalysis, dissociation and re-formation of the
thermodynamically more stable, folded RNA structure.

2.5 RNA Editing by Adenosine Deamination

RNA editing by the family of adenosine deaminases that act on RNA (ADAR
enzymes) involves the conversion of adenosine to inosine by hydrolytic N6
deamination. This modification can change RNA structure, creates new splice
sites and alters codon identities, since inosine pairs most stably with cytidine.
Editing of adenines within coding regions may therefore cause amino acid substitutions and create multiple protein isoforms from single genes. Adenosine
deamination appears to be the most widespread type of editing in higher
eukaryotes, and ADAR genes are present in most metazoan genomes (for a
review, see Bass 2002). ADAR activities were first identified in *Xenopus laevis*
as an activity that destabilizes double-stranded RNA (dsRNA) by promiscuous
deamination of adenosines within RNA duplexes (Bass and Weintraub 1988).
Since then, several cases of site-specific deamination within coding regions of
mRNAs have been reported in mammals, insects, molluscs, worms and viruses
(reviewed in Valente and Nishikura 2005).

ADAR enzymes act on RNA substrates that are either completely or largely
double-stranded. Recently, bioinformatics and experimental approaches have
revealed extensive deamination activities in humans mainly within non-coding
regions of mRNAs (reviewed in Levanon et al. 2005). Most of these A-to-I substitutions are clustered within Alu-repetitive elements that often form extended RNA
duplexes when expressed in close proximity and in opposite orientations (Fig. 2A).
Up to 50% of the As of both strands of an RNA duplex may be converted into
inosines, referred to as "hyper-editing" (DeCerbo and Carmichael 2005). A-to-I
editing is especially prevalent in the central nervous system, where it may contribute to neuronal plasticity and function. Direct sequencing of human brain cDNA
revealed an editing frequency of 1:1,000 nucleotides in non-coding regions (Blow
et al. 2004). Since the human genome contains more than a million Alu repeats that
account for ~10% of the genome, it has been suggested that most human primary
transcripts are subject to A-to-I editing (Athanasiadis et al. 2004; Levanon et al.
2004). The immediate consequences of hyper-editing are thermodynamic
destabilization of RNA duplexes (Bass and Weintraub 1988), retention of the
transcripts in the nucleus (Kumar and Carmichael 1997; Zhang and Carmichael
2001), and I-specific cleavage and degradation of RNA (Scadden and Smith 2001;
see Fig. 2A). Thus, the predominant functions of ADAR activities lie in the control
of dsRNA levels that arise from repetitive sequences and viral replication.
Furthermore, ADARs counteract dsRNA-dependent RNA interference pathways
and are likely to mediate regulatory functions of antisense transcripts in the human
genome that were recently shown to be much more prevalent than previously
thought (Katayama et al. 2005).

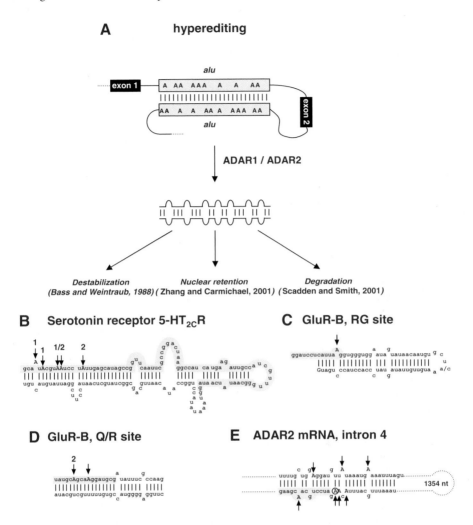

Fig. 2 RNA editing by deamination of adenines. **A** Hyper-editing of perfectly double-stranded RNA substrates frequently occurs within non-coding regions of RNA transcripts when repetitive sequences such as Alu elements are transcribed in opposite directions. ADAR1 and ADAR2 both promiscuously deaminate adenines to inosines with different cellular consequences. **B–E** Site-specific deamination within the coding regions (*grey*) of distinct mRNA targets in mammals. **B** ADAR1 and ADAR2 target sites within the serotonin 5-HT$_{2C}$R message. Several neighbouring As of an imperfect RNA duplex are edited, indicating that ADARs recognize adenines in different sequence context. **C** A single adenine is edited at the RG site of the GluR-B message. The two downstream mismatches (a/c and g/g) and the target site A/C mismatch contribute to specificity and efficiency of editing at the RG site. **D** Editing at the Q/R site is catalyzed exclusively by ADAR2. **E** ADAR2 mRNA contains a highly conserved sequence within intron 4 that folds into a conserved secondary structure. Several As within this imperfect duplex are targeted by ADAR2 enzyme, and editing at one of these positions (*circled*) creates a non-canonical splice acceptor site, leading to altered splicing and to the synthesis of a truncated and inactive version of ADAR. Thus, a negative regulatory feedback loop is created

2.5.1 A-to-I Editing Within Coding Regions

Site selective A-to-I editing in mRNA coding regions has been found in mammals, *Xenopus*, *Drosophila*, squid and several RNA viruses (reviewed in Valente and Nishikura 2005). Editing primarily affects ion channel and receptor mRNAs of the nervous systems, and often leads to functionally important changes in codon identities and amino acid sequences. Compared to the massive hyper-editing in non-coding regions described above, the few site-selective deamination events may be interpreted as "by-products" of ADAR activities that became established during evolution due to advantageous effects within a specific cell type or tissue of the organism (Maas et al. 2003).

ADAR enzymes are dsRNA binding proteins, the vertebrate family having three members, i.e. ADAR1, ADAR2 and ADAR3. ADAR1 and ADAR2 are ubiquitously expressed in many tissues, whereas ADAR3 expression is restricted to the brain (Chen et al. 2000). All ADAR enzymes have an N-terminal dsRNA binding domain with either two (ADAR2) or three (ADAR1) dsRNA binding motifs (dsRBM), whereas the C-terminal sequences encode the catalytic deaminase domain. The dsRBM is a 65-aa motif with a characteristic $\alpha\beta\beta\beta\alpha$ fold (Kharrat et al. 1995). Apart from the ADAR enzymes, the dsRBM has been found in a wide range of dsRNA binding proteins including reverse transcriptase, PKR and dicer nucleases (reviewed in Beal 2005; see Table 1). The crystal structures of the dsRBM in complex with a short RNA double strand reveal numerous contacts spanning 16 base pairs of an A-type RNA helix, including two minor grooves and the intervening major groove (Ryter and Schultz 1998; Ramos et al. 2000). Most of the contacts are sequence unspecific, involving hydrogen bonding and ionic contacts to the 2'OH groups and phospho-diesters. This explains the lack of a strict sequence requirement for dsRNA binding by the ADAR enzymes: human ADAR1 and ADAR2 both promiscuously deaminate dsRNA. Perfectly double-stranded RNAs with a minimum of 16 base pairs can function as substrates, although optimal deamination requires at least 25–30 bp and preferably >100 bp (Bass and Weintraub 1988). At the same time, ADAR1 and ADAR2 have distinct but overlapping specificities for single adenosines in only a very few transcripts. Since no auxiliary factors are needed – at least in vitro – to achieve specificity of ADARs for distinct sites, the question arises as to which factors contribute to directing the ADAR enzymes to these unique sites within the few pre-mRNA targets known to date.

2.5.2 Structural Determinants of Site Selectivity of ADAR Activity

The best characterised examples of site-selective deamination in mammals occur in transcripts of the glutamate receptor subunits gluR-B, -C and -D and gluR-5 and -6 (Fig. 2C und D), and in the transcript of the serotonin receptor subunit 2C (5-HT$_{2C}$R; Fig. 2B). Editing of several codons within these transcripts leads to altered amino acids and generates receptors with altered functions (reviewed in Seeburg and Hartner 2003). In all cases, the edited adenosines are embedded within imperfect RNA duplex structures that are formed between exon sequences and sequences with partial complementarity from downstream introns (see Fig. 2).

Structural characteristics of the exon/intron duplex as well as certain sequence preferences contribute to the selection of distinct adenosines by the ADAR enzymes. The RNA duplexes are interrupted by bulges and internal loops that were shown to separate and uncouple long RNA duplexes into shorter targets that are independently bound and modified by the ADAR enzymes (Lehmann and Bass 1999; Dawson et al. 2004). In a reciprocal approach, removal of two mismatches downstream of the R/G site in the GluR-B transcript (see Fig. 2C) did not alter the binding affinity of ADAR2 for its RNA substrate (Öhman et al. 2000). However, the editing efficiency at the R/G site was decreased due to the fact that mismatch removal allowed promiscuous ADAR binding anywhere along the resulting RNA duplex. In an extension of this study, Källman et al. (2003) could show that – although binding occurred anywhere on the RNA duplex – editing was still specific for the R/G site. This indicated that additional structural and/or sequence characteristics at or near the edited adenine contribute to editing site selection.

There is no discrete sequence motif common to all ADAR target sites and, in the case of the 5-HT$_{2C}$R mRNA editing by ADAR1 and ADAR2, five adenines in different sequence context (covering 13 nt) are edited (see Fig. 2B). However, distinct 5′-nearest neighbour preferences have been identified for ADAR1 (U = A > C > G) and ADAR2 (U = A > C = G), and the 3′ preferences for ADAR2 were U = G > C = A (Lehmann and Bass 2000). In addition, the identity of the base opposite to the editing site is crucial, with a preference for A–C mismatches or A–U base pairs, whereas editing is strongly disfavoured with an opposite purine (Källman et al. 2003). Since the editing site and the surrounding bases are contacted by the catalytic domain of the ADARs during deamination, a structural complementarity of editing site structure and the catalytic domain is required for A-site selection (Yi-Brunozzi et al. 2001). Indeed, ADAR2 binding was reported to induce conformational changes at the editing site, consistent with a model in which the adenine that is initially buried in the RNA duplex as an AU base pair or an AC mismatch is flipped out and accommodated in the catalytic centre (Stephens et al. 2000). Flipping of the adenine was dependent on the presence of the N-terminal catalytic domain, and did not occur upon dsRBM binding to the duplex (Yi-Brunozzi et al. 2001). On the contrary, dsRBM binding increased the conformational flexibility of duplex nucleotides opposite the editing site, showing that both functional domains act in concert to lower the activation energy for base flipping. The catalytic domain of ADAR2 contributes significantly to A selectivity, as demonstrated by a protein construct containing the ADAR2 catalytic domain and the ADAR1 dsRBMs: this chimeric protein retained the substrate selectivity observed for ADAR2 (Wong et al. 2001). In a recent study, sequence conservation was reported in the case of ADAR2 editing of its own mRNA that creates a non-canonical AI 3′-splice acceptor site (see Fig. 2E; Dawson et al. 2004). In addition to the 5′- and 3′-nearest neighbour preferences, seven positions within 18 nt upstream and 15 nt downstream of the editing site were conserved, and the conservation of three of these sites correlated with in vitro editing efficiency.

Taken together, site-specific adenosine deamination by the ADAR enzymes occurs by a series of dynamic interactions and structural changes. Initial binding of the dsRBMs to regions of duplexed RNA may be sequence-unspecific but these will be needed to direct the catalytic domain to the specific editing site. Sequence-specific interactions of the catalytic domain as well as unspecific binding of the dsRBMs contribute to destabilization of the local duplex and lower the activation energy for flipping the adenine out of its helical context. The adenine is accommodated in a binding pocket to allow nucleophilic attack by enzyme-bound water. The recently published crystal structure of the N-terminal half of ADAR2 supports this model, in which the adenine that is initially buried in the RNA duplex is flipped out and accommodated in the catalytic centre to allow nucleophilic attack by an enzyme-bound water molecule (Macbeth et al. 2005). However, the structure does not reveal how the catalytic domain may be engaged in sequence-specific interactions with a duplexed RNA target, and more detailed structural data will become available only from high-resolution structures of ADAR2 crystals in complex with duplex RNA.

To that end, the solution structure of an N-terminal fragment of ADAR2 encompassing the two dsRBDs was determined in complex with one of its natural substrates, the 71-nt RG site stem-loop (compare with Fig. 2C; Stefl et al. 2006). Both dsRBDs adopt the conserved αββα fold common to all dsRBD family members. While dsRBD1 interacts with the conserved pentaloop (Stefl and Allain 2005), dsRBD2 recognizes the two AC mismatches at the editing site and adjacent to it. The two bulged C-residues experienced the largest chemical shift changes upon dsRBD2 binding, indicating direct interactions with the protein. The model calculated from the NMR analyses revealed that recognition of the RNA by both RBDs is structure-specific, rather than base-specific. dsRBD1 interacts with the stem-loop structure by contacting the minor groove of the GCUCA pentaloop, and dsRBD1 recognizes the RNA helix containing two A–C mismatches separated by ten base pairs (Stefl et al. 2006).

One has to keep in mind that in vivo, ADAR enzymes compete with several nuclear RNA binding and modifying proteins that are working on any primary RNA transcript. These include hnRNP proteins that interact with any RNA transcript once it emerges from the polymerase complex. In addition, components of exon definition complexes, snRNPs, SR proteins, and dsRNA binding proteins such as drosha and dicer interact with specific sites on the RNAs. All these proteins may restrict access of the ADAR enzymes to only a few sites, and the observed target site specificity of ADAR enzymes may result rather from effective competition by these proteins for binding of ADARs to other target sites that – per se – would have a similar affinity for the ADAR enzymes. It remains to be seen whether specificity determinants in vivo rest in target site RNA structure alone and to what extent competitive binding of other proteins to potential target sites contributes to restricting access of ADARs to very few, select sites of the coding parts of the transcriptome.

2.6 Insertion/Deletion Editing in Kinetoplastids

Kinetoplastids are a group of protozoan pathogens that include *Trypanosoma* and *Leishmania* parasites responsible for severe human diseases such as African sleeping sickness and Chagas disease (Barrett et al. 2003). RNA editing in kineto-plastids is a mitochondrial process that creates functional mRNAs by the guide RNA-directed insertion and deletion of uridylate residues into most mitochondrial transcripts. It is characterised by an enzymatic reaction cycle involving the coordinated activities and interactions of gRNAs, pre-edited mRNAs and numerous protein components organized in a high molecular mass ribonucleoprotein complex, often termed the "editosome". Several purification protocols have confirmed the existence of at least 20 editing-associated proteins (see the chapter by Carnes and Stuart, this volume). Editing is initiated by the formation of a short duplex structure of 8–10 base pairs between the pre-edited mRNA and a cognate gRNA. This short duplex precisely defines the editing site: it functions as a substrate for an endoribonucleolytic activity that cleaves the pre-mRNA immediately 5′ of the helical segment. During deletion-type editing, uridylate residues are exonucleolytically removed from the 3′ end of the cleavage product, whereas insertion-type editing involves the addition of U-nucleotides. The exact number of uridylate residues to be added or deleted at each site is controlled by the gRNA sequence adjacent to the anchor region. This region, called the "information sequence", functions as a template for the removal or the addition of Us until complementarity (allowing also G:U wobble base pairs) between gRNA and pre-mRNA is achieved at each editing site. The editing cycle is completed by the ligation of the processed 5′ fragment to the 3′ fragment of the pre-mRNA.

RNA–RNA interactions play crucial roles during all steps of the editing cycle. The first step of the editing cycle is the annealing of gRNAs with pre-mRNAs via short stretches of complementarity. The efficiency of the base-paring interactions will be influenced by the RNA structures and by proteins associated with the RNAs. The secondary structure of gRNAs has been analysed by structure probing and UV melting spectroscopy of different gRNAs (Schmid et al. 1995), and a 3D model has been developed on the basis of these results (see Fig. 3B; Hermann et al. 1997). The gRNAs fold into two imperfect hairpin loops of low thermodynamic stability, with the anchor sequence identified as part of the 5′ hairpin and both terminal ends, including the poly-U-tail, being single-stranded (Fig. 3B). This higher-order tertiary structure may permit recognition by components of the editing machinery, while its low thermodynamic stability allows efficient annealing and unfolding during the editing cycle. The base-pairing interaction between gRNA and pre-mRNA was shown to be catalyzed by the RNA binding proteins gBP21 (Köller et al. 1997; Müller et al. 2001) and possibly gBP25 (Blom et al. 2001)

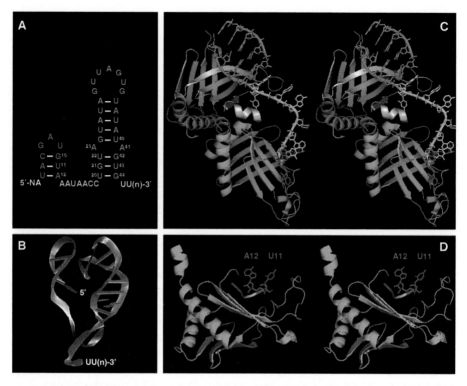

Fig. 3 Structures of gRNA alone and in complex with MRP1/2. **A** Secondary structure of gND7-506. Functional elements are colour-coded (*red* anchor sequence/stem-loop I, *green* stem-loop II, *yellow* spacer sequence). *Numbers* correspond to the nucleotides visible in the X-ray structure of the gRNA/MRP1/2 complex displayed in **C** (Schumacher et al. 2006). **B** gRNA 3D model developed by Hermann et al. (1997). **C** X-ray structure of gND7-506 in complex with MRP1 (*cyan*)/MRP2 (*pink*) heterodimer. **D** U11 and A12 (*red*) of the anchor sequence are exposed towards the solvent and are pre-oriented for the interaction with the target pre-mRNA

in *T. brucei* (also named MRP1 and MRP2; Simpson et al. 2004), and by their orthologs gbp27/29 in *C. fasciculata* (Blom et al. 2001) and Ltp26/28 in *L. tarentolae* (Aphasizhev et al. 2003a). GBP21 was shown to bind to the 3′-hairpin element of gRNAs with nanomolar affinity (Hermann et al. 1997; Köller et al. 1997). In addition, gBP21 exhibits a preference for single-stranded RNA to which it binds in a sequence-unspecific way by making use of up to six ionic contacts (Müller and Göringer 2002). On the basis of these in vitro analyses, a "matchmaker"-model for gBP21-catalyzed gRNA–pre-mRNA annealing was proposed. (1) The protein binds to the 3′ hairpin of gRNAs. (2) The 5′ end of the gRNA is unfolded in order to present the anchor region for annealing to the pre-mRNA. Up to six ionic bonds may participate in this step, thereby reducing electrostatic repulsion. (3) The anchor sequence hybridizes to the complementary target sequence on the pre-mRNA, thereby forming a partially double-stranded RNA. A 30-fold rate enhancement of gBP21-catalyzed annealing was determined with an apparent association rate constant of 3.5×10^5 M^{-1} s^{-1} (Müller et al. 2001). GBP21 was found stably associated with editing complexes (Lambert et al. 1999) where it may be engaged in protein–protein interactions – rather than binding to the editing product, since it exhibits a much lower affinity for double-stranded RNAs (Müller and Göringer 2002).

GBP21 is not essential for RNA editing, as shown by genetic knockout mutants (Lambert et al. 1999). However, it contributes to the efficiency of the editing reaction by a 30-fold acceleration of the very first step of editing that might otherwise be rate limiting at an uncatalyzed rate of 1.2×10^4 M^{-1} s^{-1}. Comparable in vitro rate enhancements were determined for a 100-kDa Ltp26/28 tetrameric complex in *L. tarentolae* (Aphasizhev et al. 2003b).

The matchmaker model of gbP21 action that was proposed on the basis of RNA modelling (Herrmann et al. 1997) and biochemical data (Müller and Göringer 2002) was essentially confirmed by the recently determined crystal structures of gBP21/25 (MRP1/MRP2) alone and in complex with gRNA (Schumacher et al. 2006). These structures revealed a novel "whirly" fold for the protein heterodimer complex (see Fig. 3C). gRNA binding occurs to a highly basic β-sheet surface present on both molecules of the protein dimer, via sequence-non-specific electrostatic contacts and hydrogen bonds. Exactly as predicted by Müller and Göringer 2002), the MRP dimer binds to the duplexed base of stem-loop II (the first four base pairs of the 3′ hairpin) while it unfolds stem-loop I and orients the first two bases of the anchor sequence towards the solvent (Fig. 3C and D). Altogether, nine positively charged amino acids are involved in gRNA binding, contacting seven backbone phosphate moieties. Specifically, the phosphate moieties of the first two anchor bases plus the adjacent nucleotide are neutralized by three arginines and a hydrogen-bonding network involving three additional amino acids. Thus, the protein dimer decreases the electrostatic repulsion between the two RNA substrates, favouring gRNA–pre-mRNA duplex formation. The two binding platforms originally proposed in the matchmaker model by Müller and Göringer (2002) can now be attributed to the two individual MRP monomers of the dimer, one (MRP2) being responsible for binding of the base of stem-loop II, while the other (MRP1) unfolds stem-loop I by binding the anchor bases and exposing them to the solvent in a conformation suitable for duplex formation (Fig. 3C and D; Schumacher et al. 2006).

Apart from gBP21 and its orthologs, two additional proteins have been described that may affect the gRNA–pre-mRNA interaction. Both the Y-box protein RBP16 (Pelletier and Read 2003) and the RGG-box protein TbRGG1 (Vanhamme et al. 1998) are able to interact with poly-U stretches present at the 3′ terminus of all gRNAs. This activity may either affect the initial annealing reaction or it may play a role during one of the later steps of the editing cycle.

The second step of editing involves the recognition of the gRNA–mRNA complex by an endonucleolytic activity that specifically cleaves the P-diester 5′ of the duplex. Recognition of the anchor duplex and the adjacent unpaired nucleotide of the pre-mRNA has to occur structure specifically, rather than sequence specifically, since hundreds of different target sites have to be processed by the endonucleolytic activities. However, cleavage efficiency is not independent of the identity of the adjacent mismatch pair (Lawson et al. 2001), implying more complex recognition and possibly contributions of editosome components other than the endonuclease(s) to cleavage site selection. The structure of the gRNA–mRNA complex resulting from the first step of editing was analysed in vitro to address the substrate structural requirements for the subsequent endonucleolytic cleavage step (Leung and Koslowsky 1999, 2001a). For three gRNA–pre-mRNA pairs, the anchor sequences were correctly positioned and base-paired. The U-tail was shown to interact with a purine-rich region upstream of and close to the editing site, whereas the information sequence was predicted to form a stem-loop between the two duplex regions. Thus, the annealing step produces gRNA–pre-mRNA duplex structures with common secondary structure features that are recognized by proteins from the editing machinery (Leung and Koslowsky 2001b). These initial studies were confirmed by solution structure probing of protein-free gRNA–pre-mRNA complexes (Yu and Koslowsky 2006). In order to gain insight into the contributions of RNA structure to each step of editing, it will be important to extend these studies by probing RNA structures in the presence of single proteins as well as editosome preparations. Thus, analysis of the structural requirements for cleavage site recognition requires the identification of the endonuclase(s) involved. Current editosome purification protocols revealed five proteins with RNase III motifs and two proteins with exonuclease motifs (Stuart et al. 2005). Three of these proteins are likely candidates for editosome endonucleases, since they contain conserved RNase III domains plus a U1-like zinc-finger and a dsRNA binding domain (KREPB1, 2 and 3). KREPB3 (kinetoplast RNA editing protein B3) was shown to be essential in vivo, and its repression inhibited cleavage of insertion editing sites in vitro (Carnes et al. 2005). Cleavage of deletion editing sites was blocked by repression of KREB1B (Trotter et al. 2005), indicating that both proteins act as endonucleases or at least are involved in endonucleolytic cleavage. However, the substrate specificity of the two candidates has not been determined, since in vitro endonuclease activity could not be shown to date.

One of the editosome components, named TbMP42, was shown to have endoribonuclease activity in vitro (Brecht et al. 2005). The protein contains two zinc-fingers and a potential OB-fold but lacks a characteristic nuclease motif. Apart from the endonuclease activity, it exhibits exoribonuclease activity with a preference for U-nucleotides. In vitro cleavage studies with recombinant TbMP42 and short model substrates revealed that the protein recognizes looped-out nucleotides with a preference for cleavage 5′ of the second mismatch nucleotide. This shift of cleavage specificity may be explained by the inaccessibility of the first mismatched nucleotide due to helical stacking on the preceding duplex in the model substrate. Thus, faithful cleavage and editing in vivo may rely on additional proteins that modify local RNA structures and help to loop out the first base for recognition by TbMP42 or by the endonuclease candidates described above.

The following steps of the editing cycle include the removal of 3′-terminal Us by exoribonuclease activities (exoUase) or the addition of Us by 3′-terminal uridylyl transferase (TUTase). Finally, RNA ligases have to join the processed ends of the pre-mRNA cleavage fragments. All enzymatic activities have been identified as parts of editing active complexes (reviewed by Stuart et al. 2005). Candidate proteins for the exoUase (Brecht et al. 2005; Kang et al. 2005), for the TUTase (Aphasizhev et al. 2003b; Ernst et al. 2003) and for the ligase(s) (Huang et al. 2001; McManus et al. 2001; Schnaufer et al. 2001) have been identified in *T. brucei* and *L. tarentolae*. Most of these proteins occur as pairs or sets with sequence similarities (Panigrahi et al. 2003), and the functions of most of the editosome proteins have been addressed by epi-genetic and/or biochemical studies using purified editosome fractions. On the basis of these studies, the existence of separate editosome complexes with different specificities for insertion or deletion editing has been proposed (Schnaufer et al. 2003; Panigrahi et al. 2006). Alternatively, the composition of the editosome may be highly dynamic, allowing the interchange of related proteins with the proposed specificities at deletion or insertion editing sites.

The candidate proteins can be classified according to their catalytic and structural motifs, such as the dsRBDs and RNase III-motifs in putative endonucleases (Worthey et al. 2003; Simpson et al. 2004; Stuart et al. 2005). However, due to the lack of in vitro activity assays based on purified proteins in most cases, little is known about the specific RNA target structures and recognition modes that are required for each individual step of the editing cycle. The presence of characteristic motifs per se is not sufficient to allow conclusions regarding protein activity, as shown for promiscuous activities (RNA binding and protein interaction) for RRMs, dsRBDs and Zn fingers (reviewed in Chen and Varani 2005). Hence, most of the RNA structural requirements have been assessed within the context of functional editosome complexes, and a number of recent studies have shed light on several general principles.

To start with, each editing site (ES) contains all the information needed to commit editing complexes to full-round insertion or deletion editing (Cifuentes-Rojas et al. 2005). An insertion ES can be converted into a deletion site by simply mutating the pre-mRNA mismatches at the ES and changing the gRNA information sequence accordingly. Hence, editing active fractions contain all enzymatic activities needed for insertion as well as deletion editing, and the ES sequence itself is the only determinant for recruitment of the necessary enzymes. While the pre-mRNA and gRNA sequences outside the ES do not affect the outcome of editing, they may influence editing efficiencies (Cifuentes-Rojas et al. 2005). It was shown that the length and thermodynamic stability of the anchor duplex directly correlate with editing efficiencies in the case of COII pre-mRNA editing directed by an engineered trans-acting gRNA (Golden and Hajduk 2006). In addition, a single gRNA point mutation in a region outside of the two duplex-forming elements flanking the editing site dramatically changes the gRNA tertiary structure and causes a strong decrease in editing rates. The mutation may inhibit a step prior to cleavage, i.e. gRNA binding/unfolding by MRP1/2 and annealing to the pre-mRNA substrate (Golden and Hajduk 2006). Conclusions will remain ambiguous, since the editing assays usually include several steps and enzymatic activities, and each of these could be rate-limiting. Therefore, more refined assays with single components are needed to dissect the RNA structural requirements for each single step of editing.

The role of RNA structure and base pairing in reactions occurring after the endonucleolytic step was addressed by a pre-cleaved editing assay (Igo et al. 2002a, 2002b). The addition of Us to the 5′-cleavage fragment by TUTase activity was not affected by the identity of the guiding nucleotide. However, the added Us are stabilized by guiding As, probably by blocking the exoUase through base pairing. Thus, base pairing determines the number of Us added and is a prerequisite for efficient ligation (see below). The position of an upstream duplex between the pre-mRNA purine stretch and the gRNA U-tail affects the efficiency of full-round editing. A 15-bp duplex immediately adjacent to an insertion site with only two A-mismatches in the gRNA was not edited, since the initial endonucleolytic cleavage step was blocked (Igo et al. 2002b). Moving the duplex at least four nucleotides further away from the ES enabled efficient cleavage, irrespective of the actual distance. This indicates that the endonuclease activity recognizes an internal loop with a minimal size and number of mismatched bases in the range of 3–6 nucleotides. However, the sum of U addition and ligation was more efficient when the duplex was moved closer to the ES, indicating that one or both of these activities prefer substrates that resemble perfectly helical duplexes. Two ligases have been identified in the *T. brucei* editosome, named TbREL1 and TbREL2 (Huang et al. 2001; McManus et al. 2001; Schnaufer et al. 2001). In vitro studies with purified and recombinant proteins have shown that both enzymes ligate RNA termini and require a bridging RNA molecule, which in vivo is performed by a gRNA (Palazzo et al. 2003). In addition, both enzymes preferentially ligate splinted RNA with no gaps, a structure that resembles a perfect RNA duplex with a nick at the ES. This preference contributes to the accuracy of editing by "counting" the correct number of Us at the ES. Unfortunately, the recently solved crystal structure of the N-terminal catalytic domain of one of the *T. brucei* ligases, TbREL1 (Deng et al. 2004), does not reveal an obvious RNA binding mode, since patches of positive electrostatic potential around the ATP binding site are missing. The only obvious determinants of RNA binding are three completely conserved aromatic F-residues that are exposed to the solvent and may be engaged in stacking interactions with RNA bases. Possibly, other editosome proteins interact with the ligases, e.g. via protein–protein interaction motifs in the C-terminal domain and, in this way, contribute substrate specificity of the *T. brucei* ligases.

Two 3′-terminal uridylyl transferases (TUTases) have been identified in *T. brucei*, one of which is part of the editosome complexes (Ernst et al. 2003). This enzyme, named TbMP57, primarily adds a single U to the 3′ end of ssRNA, with a preference for a 3′-terminal A or G. It also adds the specified number of Us to a pre-cleaved double-stranded RNA editing substrate, reflecting the characteristics of natural insertion editing events (Ernst et al. 2003). Thus, the substrate RNA structure is a major determinant of TbMP57 activity. The crystal structure of TbMP57 revealed a large, concave basic electrostatic surface that potentially interacts with the substrate RNA (Deng et al. 2005). In the absence of a co-crystallized model RNA substrate, however, predictions as to the binding modes of single-stranded or double-stranded RNA can not be made.

For deletion editing, a specified number of non-matched Us has to be deleted from the 3′ end of the 5′-cleavage fragment. Such exonuclease activities are part of the editosome and have been characterised (Aphasizhev and Simpson 2001; Igo et al. 2002b). Two candidate proteins with exonuclease motifs, TbMP100 (also: REX1) and TbMP99 (also: REX2) are part of the editosome (reviewed in Stuart et al. 2005). One of these, REX1, has been expressed as a recombinant protein and has been shown to exhibit the expected trimming activity on 3′-overhanging Us (Kang et al. 2005). A third protein, named TbMP42, which lacks any obvious exonuclease motifs was shown to have endo- and exoribonuclease activity when expressed as a recombinant protein. The protein displayed exonuclease activity in vitro with a preference for Us and also showed the specificity for 3′-overhanging single-stranded Us that is required during deletion editing (Brecht et al. 2005). The 3D structure of TbMP100/REX1 has been addressed by homology modelling (Mian et al. 2006). Apart from a potential hydrophobic pocket that would accommodate an extrahelical U-residue, no RNA binding sites were identified and thus, predictions as to the binding modes of editing substrates can not be made. Recently, reconstitution of deletion editing was achieved in a pre-cleaved editing assay with two recombinant enzymes, rREX1 (TbMP100) and rREL1 (Kang et al. 2005). Both enzymes were expressed from insect cells and, in vitro, were able to faithfully process and ligate two editing fragments in the presence of a guide RNA. Although both protein preparations were not absolutely pure, this experimental system faithfully reproduces the characteristics of editing and should allow a detailed analysis of structural requirements for exonuclease activity (rREX1) as well as ligase activity (REL1).

Taken together, a processive editosome will contain components for its structural integrity, for substrate binding, for each of the catalytic steps (cleavage, U addition or deletion, ligation) and for the translocation of the complex along a pre-edited mRNA. The shear number of components and activities points to a complex and dynamic nature of the editing reaction that resembles the dynamics of other RNA machines in the cell, the ribosome and the spliceosome. It is of greatest interest to elucidate the structure of the editosome and its RNA components. However, the isolation of structurally defined editosomes may be impossible due to their heterogeneity and dynamic nature (e.g. Panigrahi et al. 2006). Many editosome activities were found in pairs, and the existence of distinct editosomes with distinct functions in deletion or insertion editing has been suggested (reviewed in Stuart et al. 2005). Alternatively, the composition of the editosome may be dynamic, allowing the interchange of related proteins with different activities according to the specific needs of deletion or insertion editing sites.

Thus, progress in the elucidation of RNA and RNP structures will rely on more efficient purification methods and on the ability to "freeze" the editosome at distinct steps during the editing cycle. This could be achieved by protein-specific inhibitors or by antibiotics and related compounds that interact with defined RNA elements or proteins within the editing RNP complex. In this respect, combinatorial technologies may prove helpful for the isolation of high-affinity ligands and inhibitors in the form of nucleic acids (aptamers), proteins (microproteins, peptides) or small molecules.

3 Future and Conclusions

The three-dimensional structures of some of the most important players in the intracellular RNA world have become available during the last decade. High-resolution crystal structures of the small ribozymes as well as the large ribozymes, including the ribosome, have provided deep insights into the principles of RNA folding and catalysis. In addition, many RNA protein complexes involved in regulatory mechanisms and posttranscriptional processing reactions have been crystallized (as summarized in Table 1). Compared to the catalytic RNAs, the appreciation of RNA structural roles during editing processes is still in its infancy, mainly due to the lack of high-resolution structures of editing substrates and the complexes of RNA and editing factors. Elucidation of such structures may well suffer from the same inherent difficulties that also hamper spliceosome structure analysis, namely high flexibility and dynamic composition of the catalytic complexes. Already the most simple of all editing reactions, the A-to-I deamination catalyzed by the ADAR enzymes, consists of a series of events including binding, base flipping, nucleophilic attack and hydrolysis, and each step is influenced by RNA structural changes during the reaction. Thus, even with the elucidation of the three-dimensional architecture of RNA structures and RNA complexes involved in editing reactions, this will capture only frozen pictures of the molecular machines that require dynamic flexibility to accomplish their task. Even in the case of the ribosome where rapid progress from structural, biochemical, biophysical and genetic studies has been achieved, we still lack a full understanding of the mechanisms of tRNA selection and accommodation, translocation and catalysis of peptide bond formation (Korostelev et al. 2006). A detailed, static structure of editing complexes is thus only a starting point for studies that will ultimately need to explain the molecular dynamics of the different editing mechanisms.

Nevertheless, experimentally determined three-dimensional structures as well as 3D models derived from structure calculations will provide insights into the function and mechanisms of editing reactions and components. More importantly, they will inspire efforts to design new experiments that will eventually lead to an understanding of the diverse mechanisms of RNA editing in eukaryotic cells.

Acknowledgements Work in the author's laboratory is supported by the Dr. Illing Foundation.

References

Allain FH, Howe PW, Neuhaus D, Varani G (1997) Structural basis of the RNA-binding specificity of human U1A protein. EMBO J 16:5764–5772

Anant S, Davidson NO (2000) An AU-rich sequence element (UUUN[A/U]U) downstream of the edited C in apolipoprotein B mRNA is a high-affinity binding site for Apobec-1: binding of Apobec-1 to this motif in the 3' untranslated region of c-myc increases mRNA stability. Mol Cell Biol 20:1982–1992

Antson AA, Dodson EJ, Dodson G, Greaves RB, Chen X-P, Gollnick P (1999) Structure of the trp RNAbinding attenuation protein, TRAP, bound to RNA. Nature 401:235–242

Aphasizhev R, Simpson L (2001) Isolation and characterization of a U-specific 3' 5'-exonuclease from mitochondria of *Leishmania tarentolae*. J Biol Chem 276:21280–21284

Aphasizhev R, Aphasizhev I, Nelson RE, Simpson L (2003a) A 100-kD complex of two RNA-binding proteins from mitochondria of *Leishmania tarentolae* catalyzes RNA annealing and interacts with several RNA editing components. RNA 9:62–76

Aphasizhev R, Aphasizhev I, Simpson L (2003b) A tale of two TUTases. Proc Natl Acad Sci USA 100:10617–10622

Athanasiadis A, Rich A, Maas S (2004) Widespread A-to-I RNA editing of Alu containing mRNAs in the human transcriptome. PLOS Biol 2:e391

Aubourg S, Boudet N, Kreis M, Lecharny A (2000) In *Arabidopsis thaliana*, 1% of the genome codes for a novel protein family unique to plants. Plant Mol Biol 42:603–613

Ban N, Nissen P, Hansen J, Moore PB, Steitz TA (2000) The complete atomic structure of the large ribosomal subunit at 2.4 Å resolution. Science 289:905–920

Barrett MP, Burchmore RJ, Stich A, Lazzari JO, Frasch AC, Cazzulo JJ, Krishna S (2003) The trypanosomiases. Lancet 362:1469–1480

Bass BL (2002) RNA editing by adenosine deaminases that act on RNA. Annu Rev Biochem 71:817–846

Bass BL, Weintraub H (1988) An unwinding activity that covalently modifies its double-stranded RNA substrate. Cell 55:1089–1098

Battiste JL, Mao H, Rao NS, Tan R, Muhandiram DR, Kay LE, Frankel AD, Williamson JR (1996) α-Helix-RNA major groove recognition in an HIV-1 Rev peptide RRE RNA complex. Science 273:1547–1551

Beal PA (2005) Duplex RNA-binding enzymes: headliners from neurobiology, virology, and development. ChemBioChem 6:257–266

Blom D, Van den Burg J, Breek CK, Speijer D, Muijsers AO, Benne R (2001) Cloning and characterization of two guide RNA-binding proteins from mitochondria of *Crithidia fasciculata*: gBP27, a novel protein, and gBP29, the orthologue of *Trypanosoma brucei* gBP21. Nucleic Acids Res 2:2950–2962

Blow M, Futreal PA, Wooster R, Stratton MR (2004) A survey of RNA editing in human brain. Genome Res 14:2379–2387

Blum B, Bakalara N, Simpson L (1990) A model for RNA editing in kinetoplastid mitochondria: guide RNA molecules transcribed from maxicircle DNA provide the edited information. Cell 60:189–198

Brecht M, Niemann M, Schlüter E, Müller UF, Stuart K, Göringer HU (2005) TbMP42, a protein component of the RNA editing complex in African trypanosomes has endo-exoribonuclease activity. Mol Cell 17:621–630

Byrne EM, Gott JM (2002) Cotranscriptional editing of *Physarum* mitochondrial RNA requires local features of the native template. RNA 8:1174–1185

Byrne EM, Gott JM (2004) Unexpectedly complex editing patterns at dinucleotide insertion sites in *Physarum* mitochondria. Mol Cell Biol 24:7821–7828

Carnes J, Trotter JR, Ernst NL, Steinberg AG, Stuart K (2005) An essential RNAse III insertion editing endonuclease in *Trypanosoma brucei*. Proc Natl Acad Sci USA 102:16614–16619

Cermiakan N, Ikeda TM, Cedergren R, Gray MW (1996) Sequences homologous to yeast mito-
    chondrial an bacteriophage T7 and T3 RNA polymerase are widespread throughout the
    eukaryotic lineage. Nucleic Acids Res 24:648–654
Chateigner-Boutin AL, Hanson MR (2002) Cross-competition in transgenic chloroplasts express-
    ing single editing sites reveals shared cis elements. Mol Cell Biol 22:8448–8456
Chateigner-Boutin AL, Hanson MR (2003) Developmental co-variation of RNA editing extent of
    plastid editing sites exhibiting similar cis-elements. Nucleic Acids Res 31:2586–2594
Chen Y, Varani G (2005) Protein families and RNA recognition. FEBS J 272:2088–2097
Chen SH, Habib G, Yang CY, Gu ZW, Lee BR, Weng SA, Silberman SR, Cai SJ, Deslypere JP,
    Rosseneu M (1987) Apolipoprotein B-48 is the product of a messenger RNA with an organ-
    specific in-frame stop codon. Science 238:363–366
Chen CX, Cho DS, Wang Q, Lai F, Carter KC, Nishikura K (2000) A third member of the RNA-
    specific adenosine deaminase gene family, ADAR3, contains both single- and double-stranded
    RNA binding domains. RNA 6(5):755–767
Cheng YW, Visomirski-Robic LM, Gott JM (2001) Nontemplated addition of nucleotides to the
    3'-end of nascent RNA during RNA editing in *Physarum*. EMBO J 20:1405–1414
Chester A, Weinreb V, Carter CW Jr, Navaratnam N (2004) Optimization of apolipoprotein B
    mRNA editing by APOBEC1 apoenzyme and the role of its auxiliary factor, ACF. RNA
    10:1399–1411
Choury D, Araya A (2006) RNA editing site recognition in heterologous plant mitochondria. Curr
    Genet 50:405–416
Choury D, Farre JC, Jordana X, Araya A (2004) Different patterns in the recognition of editing
    sites in plant mitochondria. Nucleic Acids Res 32:6397–6406
Cifuentes-Rojas C, Halbig K, Sacharidou A, De Nova-Ocampo M, Cruz-Reyes J (2005) Minimal
    pre-mRNA substrates with natural and converted sites for full-round U insertion and U dele-
    tion RNA editing in trypanosomes. Nucleic Acids Res 33:6610–6620
Dawson TR, Sansam CL, Emeson RB (2004) Structure and sequence determinants required for
    the RNA editing of ADAR2 substrates. J Biol Chem 279:4941–4951
DeCerbo J, Carmichael GG (2005) Retention and repression: fates of hyperedited RNAs in the
    nucleus. Curr Opin Cell Biol 17:302–308
Deng J, Schnaufer A, Salavati R, Stuart KD, Hol WG (2004) High resolution crystal structure of
    a key editosome enzyme from *Trypanosoma brucei*: RNA editing ligase 1. J Mol Biol
    343:601–613
Deng J, Ernst NL, Turley S, Stuart KD, Hol WG (2005) Structural basis for UTP specificity of
    RNA editing TUTases from *Trypanosoma brucei*. EMBO J 2005 24:4007–4017
Deo RC, Bonanno JB, Sonenberg N, Burley SK (1999) Recognition of polyadenylate RNA by the
    poly(A)-binding protein. Cell 98:835–845
Ding J, Hayashi MK, Zhang Y, Manche L, Krainer AR, Xu RM (1999) Crystal structure of the
    two-RRM domain of hnRNP A1 (UP1) complexed with single-stranded telomeric DNA.
    Genes Dev 13:1102–1115
Ernst NL, Panicucci B, Igo RP Jr, Panigrahi AK, Salavati R, Stuart K (2003) TbMP57 is a 39 ter-
    minal uridylyl transferase (TUTase) of the *Trypanosoma brucei* editosome. Mol Cell
    11:1525–1536
Evans D, Marquez SM, Pace NR (2006) RNase P: interface of the RNA and protein worlds.
    Trends Biochem Sci 31:333–341
Frankel AD, Smith CA (1998) Induced folding in RNA-protein recognition: more than a simple
    molecular handshake. Cell 92:149–151
Gerber AP, Herschlag D, Brown PO (2004) Extensive association of functionally and cytotopi-
    cally related mRNAs with Puf family RNA-binding proteins in yeast. PLOS Biol 2:e79
Golden DE, Hajduk SL (2006) The importance of RNA structure in RNA editing and a potential
    proofreading mechanism for correct guide RNA:pre-mRNA binary complex formation. J Mol
    Biol 359:585–596
Gott JM (2001) RNA editing in *Physarum polycephalum*. In: Bass B (ed) RNA editing: frontiers
    in molecular biology. Oxford University Press, Oxford, UK, pp 20–37

Gott JM, Visomirski LM, Hunter JL (1993) Substitutional and insertional RNA editing of the cytochrome c oxidase subunit 1 mRNA of *Physarum polycephalum*. J Biol Chem 268:25483–25486

Gott JM, Parimi N, Bundschuh R (2005) Discovery of new genes and deletion editing in *Physarum* mitochondria enabled by a novel algorithm for finding edited mRNAs. Nucleic Acids Res 33:5063–5072

Handa N, Nureki O, Kurimoto K, Kim I, Sakamoto H, Shimura Y, Muto Y, Yokoyama S (1999) Structural basis for recognition of the tra mRNA precursor by the Sex-lethal protein. Nature 398:579–585

Hansen JL, Ippolito JA, Ban N, Nissen P, Moore PB, Steitz TA (2002) The structures of four macrolide antibiotics bound to the large ribosomal subunit. Mol Cell 10:117–128

Hausmann S, Garcin D, Delenda C, Kolakofsky D (1999a) The versatility of paramyxovirus RNA polymerase stuttering. J Virol 73:5568–5576

Hausmann S, Garcin D, Morel AS, Kolakofsky D (1999b) Two nucleotides immediately upstream of the essential A6G3 slippery sequence modulate the pattern of G insertions during Sendai virus mRNA editing. J Virol 73:343–351

Hayes ML, Reed ML, Hegeman CE, Hanson MR (2006) Sequence elements critical for efficient RNA editing of a tobacco chloroplast transcript *in vivo* and *in vitro*. Nucleic Acids Res 34:3742–3754

Hegeman CE, Hayes ML, Hanson MR (2005) Substrate and cofactor requirements for RNA editing of chloroplast transcripts in *Arabidopsis in vitro*. Plant J 42:124–132

Hermann T, Schmid B, Heumann H, Göringer HU (1997) A three-dimensional working model for a guide RNA from *Trypanosoma brucei*. Nucleic Acids Res 25:2311–2318

Holmes RK, Malim MH, Bishop KN (2007) APOBEC-mediated viral restriction: not simply editing? Trends Biochem Sci 32:118–218

Horton TL, Landweber LF (2000) Evolution of four types of RNA editing in myxomycetes. RNA 6:1339–1346

Huang CE, Cruz-Reyes J, Zhelonkina AG, O'Hearn S, Wirtz E, Sollner Webb B (2001) Roles for ligases in the RNA editing complex of *Trypanosoma brucei*: Band IV is needed for U-deletion and RNA repair. EMBO J 20:4694–4703

Hudson BP, Martinez-Yamout MA, Dyson HJ, Wright PE (2004) Recognition of the mRNA AU-rich element by the zinc finger domain of TIS11d. Nat Struct Mol Biol 11:257–264

Igo RP Jr, Lawson SD, Stuart K (2002a) RNA sequence and base pairing effects on insertion editing in *Trypanosoma brucei*. Mol Cell Biol 22:1567–1576

Igo RP Jr, Weston DS, Ernst NL, Panigrahi AK, Salavati R, Stuart K (2002b) Role of uridylate specific exoribonuclease activity in *Trypanosoma brucei* RNA editing. Eukaryot Cell 1:112–118

Johansson C, Finger LD, Trantirek L, Mueller TD, Kim S, Laird-Offringa IA, Feigon J (2004) Solution structure of the complex formed by the two N-terminal RNA-binding domains of nucleolin and a pre-rRNA target. J Mol Biol 337:799–816

Källman AM, Sahlin M, Ohman M (2003) ADAR2 A→I editing: site selectivity and editing efficiency are separate events. Nucleic Acids Res 31:4874–4881

Kambach C, Walke S, Young R, Avis JM, de la Fortelle E, Raker VA, Lührmann R, Li J, Nagai K (1999) Crystal structures of two Sm protein complexes and their implications for the assembly of the spliceosomal snRNPs. Cell 96:375–387

Kang X, Rogers K, Gao G, Falick AM, Zhou S, Simpson L (2005) Reconstitution of uridine-deletion precleaved RNA editing with two recombinant enzymes. Proc Natl Acad Sci USA 102:1017–1022

Katayama S, Tomaru Y, Kasukawa T, Waki K, Nakanishi M, Nakamura M, Nishida H, Yap CC, Suzuki M, Kawai J, Suzuki H, Carninci P, Hayashizaki Y, Wells C, Frith M, Ravasi T, Pang KC, Hallinan J, Mattick J, Hume DA, Lipovich L, Batalov S, Engstrom PG, Mizuno Y, Faghihi MA, Sandelin A, Chalk AM, Mottagui-Tabar S, Liang Z, Lenhard B, Wahlestedt C, RIKEN Genome Exploration Research Group, Genome Science Group (Genome Network Project Core Group), FANTOM Consortium (2005) Antisense transcription in the mammalian transcriptome. Science 309:1564–1566

Kharrat A, Macias MJ, Gibson TJ, Nilges M, Pastore A (1995) Structure of the dsRNA-binding domain of *E. coli* RNase III. EMBO J 14:3572–3584

Kielkopf CL, Lucke S, Green MR (2004) U2AF homology motifs: protein recognition in the RRM world. Genes Dev 18:1513–1526

Kolakofsky D, Hausmann S (1998) Cotranscriptional paramyxovirus mRNA editing: a contradiction in terms? In: Grosjean H, Benne R (eds) Modification and editing of RNA. ASM Press, Washington, DC, pp 413–420

Köller J, Müller UF, Schmid B, Missel A, Kruft V, Stuart K, Göringer HU (1997) *Trypanosoma brucei* gBP21. An arginine-rich mitochondrial protein that binds to guide RNA with high affinity. J Biol Chem 272:3749–3757

Korostelev A, Trakhanov S, Laurberg M, Noller HF (2006) Crystal structure of a 70S ribosome tRNA complex reveals functional interactions and rearrangements. Cell 126:1065–1077

Kotera E, Tasaka M, Shikanai T (2005) A pentatricopeptide repeat protein is essential for RNA editing in chloroplasts. Nature 433:326–330

Kumar M, Carmichael GG (1997) Nuclear antisense RNA induces extensive adenosine modifications and nuclear retention of target transcripts. Proc Natl Acad Sci USA 94:3542–3547

Lamb RA, Kolakofsky D (1996) Paramyxoviridae: the viruses and their replication. In: Fields BN, Knipe DM, Howley PM (eds) Fields virology, 3rd edn. Raven Press, New York, NY, pp 1177–1204

Lambert L, Müller UF, Souza AE, Göringer HU (1999) The involvement of gRNA-binding protein gBP21 in RNA editing – an *in vitro* and *in vivo* analysis. Nucleic Acids Res 27:1429–1436

Lawson SD, Igo RP Jr, Salavati R, Stuart KD (2001) The specificity of nucleotide removal during RNA editing in *Trypanosoma brucei*. RNA 7:1793–802

Lehmann KA, Bass BL (1999) The importance of internal loops within RNA substrates of ADAR1. J Mol Biol 291:1–13

Lehmann KA, Bass BL (2000) Double-stranded RNA adenosine deaminases ADAR1 and ADAR2 have overlapping specificities. Biochemistry 39:12875–12884

Leontis NB, Westhof E (2003) Analysis of RNA motifs. Curr Opin Struct Biol 13:300–308

Leontis NB, Lescoute A, Westhof E (2006) The building blocks and motifs of RNA architecture. Curr Opin Struct Biol 2006 16:279–287

Lescoute A, Leontis NB, Massire C, Westhof E (2005) Recurrent structural RNA motifs, isostericity matrices and sequence alignments. Nucleic Acids Res 33:2395–2409

Leung SS, Koslowsky DJ (1999) Mapping contacts between gRNA and mRNA in trypanosome RNA editing. Nucleic Acids Res 27:778–787

Leung SS, Koslowsky DJ (2001a) Interactions of mRNAs and gRNAs involved in trypanosome mitochondrial RNA editing: structure probing of an mRNA bound to its cognate gRNA. RNA 7:1803–1816

Leung SS, Koslowsky DJ (2001b) RNA editing in *Trypanosoma brucei*: characterization of gRNA U-tail interactions with partially edited mRNA substrates. Nucleic Acids Res 29:703–709

Levanon EY, Eisenberg E, Yelin R, Nemzer S, Hallegger M, Shemesh R, Fligelman ZY, Shoshan A, Pollock SR, Sztybel D, Olshansky M, Rechavi G, Jantsch MF (2004) Systematic identification of abundant A-to-I editing sites in the human transcriptome. Nat Biotechnol 22:1001–1005

Levanon K, Eisenberg E, Rechavi G, Levanon EY (2005) Letter from the editor. Adenosine-to inosine RNA editing in Alu repeats in the human genome. EMBO Rep 6:831–835

Lewis HA, Musunuru K, Jensen KB, Edo C, Chen H, Darnell RB, Burley SK (2000) Sequence-specific RNA binding by a nova KH domain: implications for paraneoplastic disease and the fragile X syndrome. Cell 100:323–332

Lu D, Searles MA, Klug A (2003) Crystal structure of a zinc-finger-RNA complex reveals two modes of molecular recognition. Nature 426:96–100

Ma JB, Ye K, Patel DJ (2004) Structural basis for overhang-specific small interfering RNA recognition by the PAZ domain. Nature 429:318–322

Ma JB, Yuan YR, Meister G, Pei Y, Tuschl T, Patel DJ (2005) Structural basis for 5'-end-specific recognition of guide RNA by the *A. fulgidus* Piwi protein. Nature 434:666–670

Maas S, Rich A, Nishikura K (2003) A-to-I RNA editing: recent news and residual mysteries. J Biol Chem 278:1391–1394

Macbeth MR, Schubert HL, Vandemark AP, Lingam AT, Hill CP, Bass BL (2005) Inositol hexakisphosphate is bound in the ADAR2 core and required for RNA editing. Science 309:1534–1539

Macdonald LE, Zhou Y, McAllister WT (1993) Termination and slippage by bacteriophage T7 RNA polymerase. J Mol Biol 232:1030–1047

MacRae IJ, Zhou K, Li F, Repic A, Brooks AN, Cande WZ, Adams PD, Doudna JA (2006) Structural basis for double-stranded RNA processing by Dicer. Science 311:195–198

Maier RM, Zeltz P, Kossel H, Bonnard G, Gualberto JM, Grienenberger JM (1996) RNA editing in plant mitochondria and chloroplasts. Plant Mol Biol 32:343–365

Maris C, Masse J, Chester A, Navaratnam N, Allain FH (2005a) NMR structure of the apoB mRNA stem-loop and its interaction with the C to U editing APOBEC1 complementary factor. RNA 11:173–186

Maris C, Dominguez C, Allain FH (2005b) The RNA recognition motif, a plastic RNA-binding platform to regulate post-transcriptional gene expression. FEBS J 272:2118–2131

Masters BS, Stohl LL, Clayton AD (1987) Yeast mitochondrial RNA polymerase is homologous to those encoded by bacteriophages T3 and T7. Cell 51:98–99

McManus MT, Shimamura M, Grams J, Hajduk SL (2001) Identification of candidate mitochondrial RNA editing ligases from *Trypanosoma brucei*. RNA 7:167–175

Mehta A, Driscoll DM (2002) Identification of domains in apobec-1 complementation factor required for RNA binding and apolipoprotein-B mRNA editing. RNA 8:69–82

Mehta A, Kinter MT, Sherman NE, Driscoll DM (2000) Molecular cloning of apobec-1 complementation factor, a novel RNA-binding protein involved in the editing of apolipoprotein B mRNA. Mol Cell Biol 20:1846–1854

Mian IS, Worthey EA, Salavati R (2006) Taking U out, with two nucleases? BMC Bioinformatics 16:305

Miller ML, Antes TJ, Qian F, Miller DL (2006) Identification of a putative mitochondrial RNA polymerase from *Physarum polycephalum*: characterization, expression, purification, and transcription *in vitro*. Curr Genet 49:259–271

Milligan JF, Groebe DR, Witherell GW, Uhlenbeck OC (1987) Oligoribonucleotide synthesis using T7 RNA polymerase and synthetic DNA templates. Nucleic Acids Res 15:8783–8798

Miyamoto T, Obokata J, Sugiura M (2002) Recognition of RNA editing sites is directed by unique proteins in chloroplasts: biochemical identification of cis-acting elements and trans-acting factors involved in RNA editing in tobacco and pea chloroplasts. Mol Cell Biol 22:6726–6734

Miyamoto T, Obokata J, Sugiura M (2004) A site-specific factor interacts directly with its cognate RNA editing site in chloroplast transcripts. Proc Natl Acad Sci USA 101:48–52

Moore MJ (2005) From birth to death: the complex lives of eukaryotic mRNAs. Science 309:1514–1518

Moore PB, Steitz TA (2003) The structural basis of large ribosomal subunit function. Annu Rev Biochem 72:813–850

Müller UF, Göringer HU (2002) Mechanism of the gBP21-mediated RNA/RNA annealing reaction: matchmaking and charge reduction. Nucleic Acids Res 30:447–455

Müller UF, Lambert L, Göringer HU (2001) Annealing of RNA editing substrates facilitated by guide RNA-binding protein gBP21. EMBO J 20:1394–1404

Müller M, Heuck A, Niessing D (2007) Directional mRNA transport in eukaryotes: lessons from yeast. Cell Mol Life Sci 64:171–180

Nakamura T, Meierhoff K, Westhoff P, Schuster G (2003) RNA-binding properties of HCF152, an *Arabidopsis* PPR protein involved in the processing of chloroplast RNA. Eur J Biochem 270:4070–4081

Nelson JA, Uhlenbeck OC (2006) When to believe what you see. Mol Cell 23:447–450

Noller HF (2005) RNA structure: reading the ribosome. Science 309:1508–1514

Öhman M, Kallman AM, Bass BL (2000) *In vitro* analysis of the binding of ADAR2 to the pre-mRNA encoding the GluR-B R/G site. RNA 6:687–697

Palazzo SS, Panigrahi AK, Igo RP, Salavati R, Stuart K (2003) Kinetoplastid RNA editing ligases: complex association, characterization, and substrate requirements. Mol Biochem Parasitol 127:161–167

Panigrahi AK, Schnaufer A, Ernst NL, Wang B, Carmean N, Salavati R, Stuart K (2003) Identification of novel components of *Trypanosoma brucei* editosomes. RNA 9:484–492

Panigrahi AK, Ernst NL, Domingo GJ, Fleck M, Salavati R, Stuart KD (2006) Compositionally and functionally distinct editosomes in *Trypanosoma brucei*. RNA 12:1038–1049

Parker JS, Roe SM, Barford D (2005) Structural insights into mRNA recognition from a PIWI domain-siRNA guide complex. Nature 434:663–666

Pelletier M, Read LK (2003) RBP16 is a multifunctional gene regulatory protein involved in editing and stabilization of specific mitochondrial mRNAs in *Trypanosoma brucei*. RNA 9:457–468

Powell LM, Wallis SC, Pease RJ, Edwards YH, Knott TJ, Scott J (1987) A novel form of tissue-specific RNA processing produces apolipoprotein-B48 in intestine. Cell 50:831–840

Price SR, Evans PR, Nagai K (1998) Crystal structure of the spliceosomal U2B''-U2A' protein complex bound to a fragment of U2 small nuclear RNA. Nature 394:645–650

Puglisi JD, Chen L, Blanchard S, Frankel AD (1995) Solution structure of a bovine immunodeficiency virus Tat-TAR peptide-RNA complex. Science 270:1200–1203

Ramos A, Grunert S, Adams J, Micklem DR, Proctor MR, Freund S, Bycroft M, St Johnston D, Varani G (2000) RNA recognition by a Staufen double stranded RNA binding domain. EMBO J 19:997–1009

Reichow SL, Hamma T, Ferre-D'Amare AR, Varani G (2007) The structure and function of small nucleolar ribonucleoproteins. Nucleic Acids Res 35:1452–1464

Richardson N, Navaratnam N, Scott J (1998) Secondary structure for the apolipoprotein B mRNA editing site. Au-binding proteins interact with a stem loop. J Biol Chem 273:31707–31717

Rigaut G, Shevchenko A, Rutz B, Wilm M, Mann M, Seraphin B (1999) A generic protein purification method for protein complex characterization and proteome exploration. Nat Biotechnol 17:1030–1032

Ryter JM, Schultz SC (1998) Molecular basis of double-stranded RNA-protein interactions: structure of a dsRNA-binding domain complexed with dsRNA. EMBO J 17:7505–7513

Scadden AD, Smith CW (2001) Specific cleavage of hyper-edited dsRNAs. EMBO J 20:4243–4252

Schmid B, Riley GR, Stuart K, Göringer HU (1995) The secondary structure of guide RNA molecules from *Trypanosoma brucei*. Nucleic Acids Res 23:3093–3102

Schnaufer A, Panigrahi AK, Panicucci B, Igo R Jr, Salavati R, Stuart K (2001) An RNA ligase essential for RNA editing and survival of the bloodstream form of *Trypanosoma brucei*. Science 291:2159–2162

Schnaufer A, Ernst N, O'Rear J, Salavati R, Stuart K (2003) Separate insertion and deletion sub-complexes of the *Trypanosoma brucei* RNA editing complex. Mol Cell 12:307–319

Schumacher MA, Karamooz E, Zikova A, Trantirek L, Lukes J (2006) Crystal structures of *T. brucei* MRP1/MRP2 guide-RNA binding complex reveal RNA matchmaking mechanism. Cell 126:701–711

Seeburg PH, Hartner J (2003) Regulation of ion channel/neurotransmitter receptor function by RNA editing. Curr Opin Neurobiol 13:279–283

Simpson L, Aphasizhev R, Gao G, Kang X (2004) Mitochondrial proteins and complexes in *Leishmania* and *Trypanosoma* involved in U insertion/deletion RNA editing. RNA 10:159–170

Small ID, Peeters N (2000) The PPR motif – a TPR-related motif prevalent in plant organellar proteins. Trends Biochem Sci 25:46–47

Song JJ, Smith SK, Hannon GJ, Joshua-Tor L (2004) Crystal structure of Argonaute and its implications for RISC slicer activity. Science 305:1434–1437

Stark H, Lührmann R (2006) Cryo-electron microscopy of spliceosomal components. Annu Rev Biophys Biomol Struct 35:435–457

Stefl R, Allain FH (2005) A novel RNA pentaloop fold involved in targeting ADAR2. RNA 11:592–597

Stefl R, Xu M, Skrisovska L, Emeson RB, Allain FH (2006) Structure and specific RNA binding of ADAR2 double-stranded RNA binding motifs. Structure 14(2):345–355

Stephens OM, Yi-Brunozzi HY, Beal PA (2000) Analysis of the RNA editing reaction of ADAR2 with structural and fluorescent analogues of the GluR-B R/G editing site. Biochemistry 39:12243–12251

Stuart KD, Schnaufer A, Ernst NL, Panigrahi AK (2005) Complex management: RNA editing in trypanosomes. Trends Biochem Sci 30:97–105

Takenaka M, Neuwirt J, Brennicke A (2004) Complex cis-elements determine an RNA editing site in pea mitochondria. Nucleic Acids Res 32:4137–4144

Teng B, Burant CF, Davidson NO (1993) Molecular cloning of an apolipoprotein B messenger RNA editing protein. Science 260:1816–1819

Tillich M, Funk HT, Schmitz-Linneweber C, Poltnigg P, Sabater B, Martin M, Maier RM (2005) Editing of plastid RNA in *Arabidopsis thaliana* ecotypes. Plant J 43:708–715

Trotter JR, Ernst NL, Carnes J, Panicucci B, Stuart K (2005) A deletion site editing endonuclease in *Trypanosoma brucei*. Mol Cell 20:403–412

Tsudzuki T, Wakasugi T, Sugiura M (2001) Comparative analysis of RNA editing sites in higher plant chloroplasts. J Mol Evol 53:327–332

Turelli P, Trono D (2005) Editing at the crossroad of innate and adaptive immunity. Science 307:1061–1065

Valadkhan S (2005) snRNAs as the catalysts of pre-mRNA splicing. Curr Opin Chem Biol 9:603–608

Valencia-Sanchez MA, Liu J, Hannon GJ, Parker R (2006) Control of translation and mRNA degradation by miRNAs and siRNAs. Genes Dev 20:515–524

Valente L, Nishikura K (2005) ADAR gene family and A-to-I RNA editing: diverse roles in posttranscriptional gene regulation. Prog Nucleic Acid Res Mol Biol 79:299–338

van der Merwe JA, Takenaka M, Neuwirt J, Verbitskiy D, Brennicke A (2006) RNA editing sites in plant mitochondria can share cis-elements. FEBS Lett 580:268–272

Vanhamme L, Perez-Morga D, Marchal C, Speijer D, Lambert L, Geuskens M, Alexandre S, Ismaili N, Göringer U, Benne R, Pays E (1998) *Trypanosoma brucei* TBRGG1, a mitochondrial oligo(U)-binding protein that co-localizes with an *in vitro* RNA editing activity. J Biol Chem 273:21825–21833

Vicens Q, Cech TR (2006) Atomic level architecture of group I introns revealed. Trends Biochem Sci 31:41–51

Vidal S, Curran J, Kolakofsky D (1990) A stuttering model for paramyxovirus P mRNA editing. EMBO J 9:2017–2022

Wang SS, Mahendran R, Miller DL (1999) Editing of cytochrome b mRNA in *Physarum* mitochondria. J Biol Chem 274:2725–2731

Wang X, McLachlan J, Zamore PD, Hall TM (2002) Modular recognition of RNA by a human pumilio-homology domain. Cell 110:501–512

Wimberly BT, Brodersen DE, Clemons WM Jr, Morgan-Warren RJ, Carter AP, Vonrhein C, Hartsch T, Ramakrishnan V (2000) Structure of the 30S ribosomal subunit. Nature 407:327–339

Wong SK, Sato S, Lazinski DW (2001) Substrate recognition by ADAR1 and ADAR2. RNA 7:846–858

Worthey EA, Schnaufer A, Mian IS, Stuart K, Salavati R (2003) Comparative analysis of editosome proteins in trypanosomatids. Nucleic Acids Res 31:6392–6408

Xie K, Sowden MP, Dance GS, Torelli AT, Smith HC, Wedekind JE (2004) The structure of a yeast RNA-editing deaminase provides insight into the fold and function of activation induced deaminase and APOBEC-1. Proc Natl Acad Sci USA 101:8114–8119

Ye X, Kumar RA, Patel DJ (1995) Molecular recognition in the bovine immunodeficiency virus Tat peptide–TAR RNA complex. Chem Biol 2:827–840

Yi-Brunozzi HY, Stephens OM, Beal PA (2001) Conformational changes that occur during an RNA-editing adenosine deamination reaction. J Biol Chem 276:37827–37833

Yu LE, Koslowsky DJ (2006) Interactions of mRNAs and gRNAs involved in trypanosome mitochondrial RNA editing: structure probing of a gRNA bound to its cognate mRNA. RNA 12:1050–1060

Zhang Z, Carmichael GG (2001) The fate of dsRNA in the nucleus: a p54(nrb)-containing complex mediates the nuclear retention of promiscuously A-to-I edited RNAs. Cell 106:465–475

Editing of tRNA for Structure and Function

Juan D. Alfonzo

"The difference between the right word and the almost right word is the difference between lightning and the lightning bug."

Mark Twain
On editing.

Abstract The degeneracy of the genetic code is implied in the need for 61 sense codons to specify 20 different amino acids; with the exception of methionine and tryptophan, each amino acid is encoded by more than one codon. This discrepancy between codon and amino acid numbers was first explained by Crick's wobble hypothesis, which invoked the need for base-pairing flexibility between the first anticodon and third codon positions during decoding. Since the inception of the wobble rules, over 100 posttranscriptional modifications have been described, with the largest number affecting the anticodon of tRNA. As anticodon modifications accrue, new findings lead to a constant reinterpretation of the wobble rules to include novel effects on tRNA function. In general, anticodon modifications play key roles in translational fidelity and efficiency. However, anticodon-sequence alterations to a particular tRNA that permit decoding of multiple codons are part of a growing number of posttranscriptional changes collectively known as tRNA editing. In fact, the decoding changes imparted by tRNA editing provide a mechanism to effectively accommodate genetic code degeneracy. Although a

Department of Microbiology, and The Ohio State Biochemistry Program, The Ohio State University, 484 West 12th Avenue, Columbus, Ohio 43210, USA; alfonzo.1@osu.edu

H.U. Göringer (ed.), *RNA Editing. Nucleic Acids and Molecular Biology 20*
© Springer-Verlag Berlin Heidelberg 2008

number of editing events have direct effects in expanding a tRNA's decoding capacity, some editing events indirectly affect tRNA function by repairing otherwise non-functional tRNAs. This chapter will attempt to summarize what is currently known about both types of tRNA editing in various organisms, with the proviso that due to the serendipitous nature of editing discoveries, the work presented here will undoubtedly not be conclusive. This chapter will rather compile the few existing examples of tRNA editing, and whenever possible will try to illustrate current efforts to characterize the different tRNA editing enzymes and the various mechanisms.

1 Introduction – tRNA Editing: an Evolving Concept

"RNA editing", as originally stated (and discussed elsewhere in this volume), refers to the posttranscriptional alteration of sequence information in mRNA beyond what is encoded in the DNA genome from various organisms (Benne et al. 1986). Initially, this definition was sufficient to explain the mechanism of insertion and deletion of nucleotides into the pre-mRNAs of trypanosomatid mitochondria, as well as the single nucleotide changes to the coding regions of mammalian mRNAs. However, soon after the discovery of mRNA editing, the report of similar nucleotide changes in non-coding RNAs (mainly tRNAs) required the use of a broader term. Gray and co-workers first reported the posttranscriptional substitution of nucleotides to a non-coding RNA (Lonergan and Gray 1993). They found that nucleotides were added to the acceptor stem of *A. castellani* tRNAs as a required step in their maturation. They then expanded the definition of editing to include any alteration in the sequence of an RNA (coding or non-coding) that leads to the introduction of one of the four canonical nucleotides (Covello and Gray 1998). According to Covello and Gray, such "programmed alterations" result in the generation of transcripts of which the sequence could have been potentially encoded in the DNA genome. Most recently, in an attempt to differentiate between RNA editing and modification, Grosjean has adopted a more strict definition of editing whereby, regardless of the mechanism, any sequence alteration that changes the genetic meaning of a transcript is called editing, while merely structural changes are called modification (Grosjean and Björk 2004).

Arguably, changes in structural information also represent a form of programmed alteration of genetic meaning, but in this chapter, rather than trying to establish arguments for or against either definition, we will use both. Thus, this chapter will divide tRNA editing into two major groups: functional editing (using Grosjean's definition), and structural editing (using Gray's definition). To this end, this chapter will summarize what is currently known about tRNA editing changes that have direct bearings on a tRNA's function by altering its identity (i.e., expanding its decoding properties), as strictly defined by Grosjean. These include A to I, C to U, and lysidine formation at the three-anticodon nucleotides of tRNA. This chapter, however, will also include a number of canonical nucleotide changes occurring posttranscriptionally that repair and restore tRNA structure (indirectly affecting function), and that have been described in the literature as tRNA editing based on the broader definition originally coined by Covello and Gray.

2 Editing of tRNAs by Adenosine to Inosine Conversion

The discovery of the nucleoside inosine (Holley et al. 1965a, b), by Holley and co-workers over 40 years ago, caused an immediate and now historical stir. Scientist at the time were amidst trying to explain how (or why) 20 different amino acids required 61 different codons for protein synthesis. The discovery of inosine in yeast tRNA[Ala] (AUA) (because of its predicted base-pairing properties) led Crick to propose that the base-pair capabilities of this tRNA were not limited to the four canonical nucleotides. Crick reasoned that the newly discovered, inosine-containing tRNA[Ala] (AUA) could pair with three different codons (ending in A, C, or U) to specify the same amino acid. Crick further elaborated this notion into his now famous "wobble hypothesis" (Crick 1966), which provides explanation for (1) the presence of multiple codons for one amino acid, (2) the existence of more than one tRNA to specify a single amino acid, and (3) the implicit ability of a single tRNA to decode more than one codon (as in the case of inosine). It is now well established that tRNAs encoded with an adenosine at position 34 are almost universally edited to inosine, expanding the decoding capacity of a given tRNA (Fig. 1A).

Despite the early discovery of inosine, it was the pioneering work of Grosjean and co-workers 30 years later that led to the identification of an enzymatic activity in S100 supernatants that could convert adenosine to inosine in tRNA (Auxilien et al. 1996). They showed that, in vitro, extracts from yeast, *Xenopus*, mammalian cells, and bacteria could specifically convert adenosine to inosine at the first anti-codon position in tRNAs. These authors demonstrated that incubation of a tRNA substrate labeled at either the phosphate backbone or the base led to preservation of the label following inosine formation, ruling out possible mechanisms that involve complete replacement of the nucleotide (insertion/deletion-type mechanisms) or base replacement (transglycosylation-type mechanisms; Grosjean et al. 1996a). Furthermore, when unlabeled substrate was incubated in a medium of S100 supernatants and [O^{18}] water, incorporation of labeled oxygen was observed concomitantly with the formation of inosine (Auxilien et al. 1996; Grosjean et al. 1996a). These experiments, of course, were guided by similar experiments performed earlier to unveil the mechanism of mRNA editing in *C. elegans* and mammalian cells (as discussed elsewhere in this volume). Together, these findings led to the conclusion that inosine formation, regardless of the substrate (mRNA or tRNA), proceeded by a hydrolytic adenosine to inosine deamination mechanism (Fig. 1B). To date, however, only the yeast and bacterial enzymes have been characterized to some extent.

A

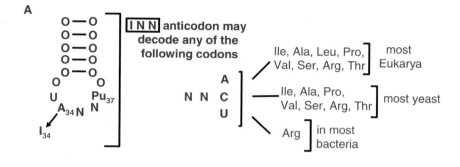

B

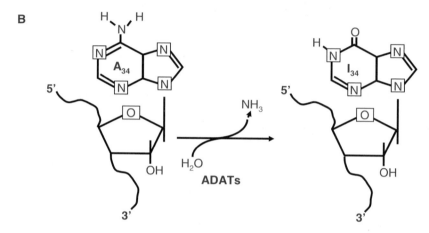

Fig. 1 Inosine at the first position of the anticodon: significance and mechanism. **A** A34 is almost universally changed to inosine in most organisms, and is necessary for expanding the decoding capacity of a tRNA to decode multiple codons by wobbling. *INN* Inosine-containing anticodon (*N* any of the four canonical nucleotides); *arrow* denotes the position of the A to I editing event in tRNA. **B** Mechanism of A to I editing in tRNAs, where the editing enzymes mediate the hydrolytic deamination of A34 to form inosine. *5'* and *3'* Corresponding ends of the tRNA, *ADATs* adenosine deaminases acting on tRNA (according to accepted nomenclature)

Several years later, knowledge of a conserved deamination mechanism prompted Gerber and Keller to search the then newly sequenced yeast genome for open reading frames encoding putative deaminases. They found that in yeast the enzyme responsible for A to I conversion at the wobble position had two subunits, the products of the ADAT2 and ADAT3 genes (for adenosine deaminase acting on tRNA, using the accepted editing nomenclature, and replacing their former names of tad2p and tad3p; Gerber and Keller 1999). Although both proteins are homologous, they are not identical, differing both in sequence and size. ADAT2 is a 24-kDa molecular mass protein that harbors all the conserved motifs required for deamination (Fig. 2). ADAT3, by contrast, is a 30-kDa molecular mass polypeptide that contains a conserved zinc-binding motif, but lacks a highly conserved proton-shuttling domain

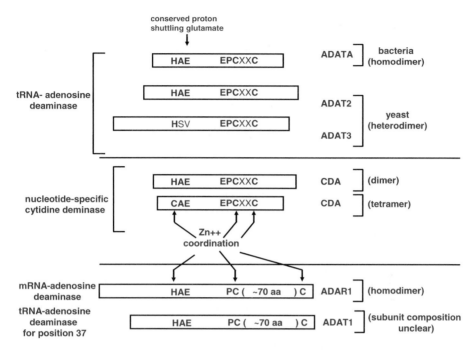

Fig. 2 Different tRNA deaminases compared to their nucleotide- and mRNA-specific counterparts. *Arrows* denote the conserved glutamate (*E*) involved in proton shuttling, and the conserved H/C and PCxxC motifs involved in Zn++ coordination, both being essential for activity

found in most deaminases (Fig. 2). Neither of the subunits, however, is by itself able to support editing of tRNAs (Gerber and Keller 1999). Therefore, in yeast, the editing enzyme is formed by heterodimerization of two subunits, which upon association, create a functional enzyme that can then specifically deaminate the wobble base of all seven A34-containing tRNAs in this organism (Gerber and Keller 1999). Interestingly, although these enzymes perform A to I editing in vivo and in vitro, their core sequences have the signatures of cytidine deaminases (rather than adenosine deaminases; Fig. 2). This has led Keller and co-workers to propose an evolutionary path for these enzymes whereby they would be derived from a gene duplication of a cytidine deaminase ancestor. This proposal thus suggests that through evolution, changes accumulated in the duplicated gene that led to the conversion of a C to U deaminase into an A to I- specific enzyme. This is in sharp contrast with the only other known tRNA deaminase, ADAT1, which forms inosine at position 37 in eukaryotic tRNAs. ADAT1 contains a set of conserved core sequences that resembles those of adenosine deaminase acting on RNA ADAR1 (an mRNA editing enzyme), and presumably appeared by a different evolutionary path. To date, it is not clear what features of the individual proteins by themselves (in the case of *E. coli*, see below), or upon association of two different subunits (in the case of yeast) change the enzyme specificity from a pyrimidine to a purine deaminase.

Our laboratory has been studying A to I editing of tRNAs in trypanosomatids (*Trypanosoma* and *Leishmania*); like in mammals, eight different tRNAs containing an A at the first position of the anticodon can be identified by genomic database searches. These tRNAs undergo A to I editing at position 34 of the anticodon (Rubio and Alfonzo, unpublished data). This editing is essential to decode the C-ending codons for the amino acids isoleucine (Ile), alanine (Ala), leucine (Leu), proline (Pro), valine (Val), serine (Ser), arginine (Arg), and threonine (Thr). We have investigated the A to I editing of threonyl tRNA (tRNA^{Thr}) based on the ease of folding of synthetic versions of this tRNA. Under in vitro conditions, synthetic tRNA^{Thr} is efficiently aminoacylated with crude synthetase fractions from *Leishmania* and *Trypanosoma* (Alfonzo and Ibba, unpublished data). In *Trypanosoma brucei*, there are four codons and three genes encoding iso-accepting tRNAs for the amino acid threonine. Two of these iso-acceptors (anticodon CGU and UGU) decode the ACG and AGA, respectively. The remaining tRNA (anticodon AGU) can decode the ACU codon, but is unable to decode the ACC codon. No tRNA is encoded in the *T. brucei* genome that may decode the ACC codon. The tRNA^{Thr} AGU must therefore undergo A to I$_{34}$ editing to expand its decoding capacity. Unexpectedly, we found that this tRNA also undergoes C to U conversion at position 32 of the same anticodon loop (Rubio and Alfonzo, unpublished data). Currently, it is not clear what is the biological significance of the two editing events. Nevertheless, in vitro we could show that C to U formation at position 32 has a stimulatory effect in the further conversion of A to I at the wobble base. Additionally, both events appear to occur outside the organelle, indicating that tRNA editing by C to U conversion is more widespread than previously thought (Rubio, Ragone and Alfonzo, unpublished data).

In bacteria, only tRNA^{Arg} contains inosine at the first anticodon position. The *E. coli* enzyme, the product of the adatA gene, is similar to the smaller subunit of the yeast enzyme (Wolf et al. 2002), and catalyzes the same A to I editing, but unlike the yeast enzyme, it can deaminate much smaller substrates, including molecules that are essentially short versions of the anticodon stem-loop (ASL; Grosjean et al. 1996b; Gerber and Keller 1999; Wolf et al. 2002). This enzyme is very specific in that it is able to deaminate the cognate bacterial tRNA^{Arg}, but is unable to edit any of the eukaryotic A34-containing tRNAs (Wolf et al. 2002). However, the ADATa protein efficiently deaminates eukaryotic tRNAs that contain a transposed bacterial tRNA^{Arg} ASL. Also unlike yeast, and presumably other eukaryotes, ADATa forms a functional, albeit weak, homodimer in solution, which has been confirmed in the recently solved structure of the *Aquifex aeolicus* enzyme (Kuratani et al. 2005). This shows that the two subunits form a three-layered $\alpha/\beta/\alpha$ structure, with a dimerization interface so extensive that it buries over 16% of the total monomer surface area (or 1,300 Å^2 of a total of 8,100 Å^2). Interestingly, an Asp to Glu change at a highly conserved residue within the dimerization interface of the bacterial proteins yields an enzyme that is inactive in vitro, but fully active in vivo. This suggests that this region is either important to stabilize the enzyme, or is needed in vivo to interact with other proteins, which then leads to structural stability (Kuratani et al. 2005). This latter observation is of special interest, given the finding by Grosjean and co-workers that the native mammalian enzyme is of a much larger size (>200 kDa; Auxilien et al. 1996) than that expected through homo- or heterodimerization of two putative subunits. This suggests that within cells, the A to I tRNA editing enzyme is part of a much larger protein complex.

Although the enzymes involved in A to I editing of tRNA have been identified in yeast and bacteria, and in vitro assays exist for a number of organisms (including *Xenopus*, trypanosomatids, and humans), these activities are far from being fully characterized. To date, it is not known how these enzymes specifically deaminate the first anticodon position, or what is the basis for tRNA binding. From an evolutionary and mechanistic standpoint, it is also unclear what events led to the conversion of cytidine deaminases into adenosine deaminases.

3   C to U Editing of tRNAs

Unlike its A to I counterpart, to date there are only two examples of C to U editing of tRNA at either of the three anticodon nucleotides: C to U editing of tRNA^{Gly} in marsupial mitochondria (Janke and Pääbo 1993), which affects the second position of the anticodon (C35), and C to U editing of tRNA^{Trp} in trypanosomatid mitochondria, specific for the first position (C34; Alfonzo et al. 1999; Figs. 3 and 4).

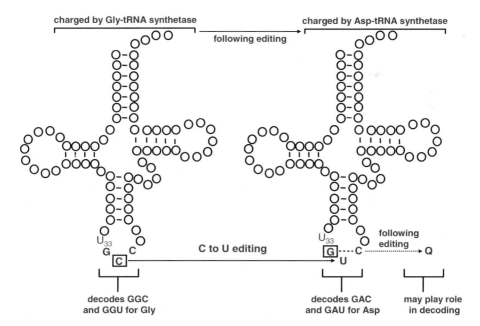

Fig. 3 C to U editing in marsupial mitochondria. This editing event changes the amino acid and the decoding capacity of the tRNA, and also affects the formation of queuosine (*Q*). The *arrows* indicate the various changes and their consequences

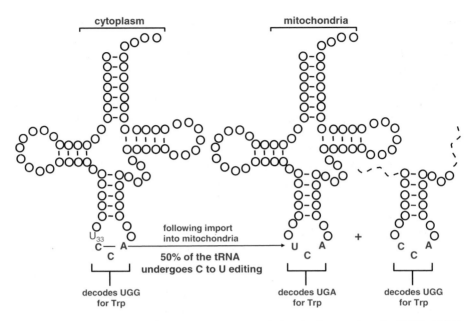

Fig. 4 C to U editing in trypanosomatid mitochondria. Following import, the only tRNA^{Trp} (CCA) encoded in the genome is imported into the mitochondria where it is edited. This editing is thought to be essential for decoding the mitochondrial UGA codons as tryptophan

In marsupial mitochondria, Janke and Pääbo first observed the absence of a gene for a tRNA that could decode the mitochondrial aspartate codons. They found that at the position of the mitochondrial genome where most mammals encode for tRNA^{Asp}, the gene had been replaced by a tRNA^{Gly}GCC. This tRNA could predictably decode two of four mitochondrial codons for glycine (i.e., GGY), but would fail to decode the Asp codons. Sequence comparison with tRNA databases showed that, except for the three-anticodon nucleotides, the tRNA^{Gly} was very similar to tRNA^{Asp} from other organisms (Janke and Pääbo 1993). These researchers suggested, and later demonstrated, that this tRNA^{Gly} undergoes a single cytosine to uridine (C to U) editing at the second position of the anticodon (C35), which changes the codon specificity of this tRNA from glycine to aspartate (Janke and Pääbo 1993; Janke et al. 1994; Mörl et al. 1995; Börner et al. 1996; Fig. 3). They also showed that this editing event was widespread in marsupial mitochondria isolated from different tissues, but that the editing levels were fairly constant among tissues. Thus, editing does not appear to play a role in tissue-specific gene expression. Interestingly, only 50% of the tRNA^{Gly}GCC was converted to tRNA^{Asp} by editing (Mörl et al. 1995), raising the possibility that either both the unedited and edited tRNAs play a role in mitochondrial translation, or only the edited species is functional but editing is inefficient. Further testing revealed that in vitro transcripts representing the two tRNA versions could be efficiently charged with synthetase fractions, but the tRNA^{Gly} could be charged only with glycine while the edited tRNA could accept only aspartate (Börner et al. 1996). Use of an elegantly designed assay (OXOCIRC), where the ability of sodium periodate to oxidized adjacent free hydroxyl (i.e., those found at the 3′ of RNA) to form dialdehydes was exploited, led to the assessment of whether or not both tRNAs were also charged in vivo (Börner et al. 1996). In this assay, the presence of an amino acid at the 3′ end of the tRNA blocks the accessibility of the end to oxidation, and when coupled with circularization by RNA ligase and RT-PCR, permits quantitation of the number of charged tRNAs, regardless of their anticodon-sequence identity. These studies led to the conclusion that both the unedited and edited versions of the tRNA were functional. However, it is not clear what determines the 50% balance in terms of C to U conversion. Furthermore, the biological significance of this balance, although also unclear, perhaps suggests some function in decoding.

Besides affecting charging and codon recognition, C to U editing of tRNA^{Gly} also plays a role as a modification determinant. The Pääbo laboratory demonstrated that the C to U conversion at position 35 also led to the creation of the sequence motif (UGU, where the last U is created by C to U editing) needed for further conversion of G34 into the hypermodified nucleoside queosine (Q; Mörl et al. 1995). The nucleoside Q, in turn, has been implicated in affecting in vivo decoding by apparently stabilizing codon–anticodon interactions (Fig. 3; Urbonavicius et al. 2001, 2003).

We discovered the only other example of C to U editing in the anticodon of a tRNA, which occurs in the mitochondria of trypanosomatids (Alfonzo et al. 1999). In these organisms, the mitochondrial genome does not encode a single tRNA gene. Thus, for mitochondrial translation to take place, every tRNA is synthesized in the nucleus, transits through the cytoplasm, and is then imported into the mitochondria. The single nucleus-encoded tryptophanyl tRNA (tRNA^{Trp}) is transcribed with a CCA anticodon, posing the conundrum of how this tRNA decodes mitochondrial UGA codons. In trypanosomatid mitochondria, like in most eukaryotes, the UGA codons have been reassigned to tryptophan, while in the cytoplasm UGA functions as a stop codon (opal codon). Trypanosomatids need to decode the UGA codons, and concurrently prevent suppression of the same codon while the tRNA transits through the cytoplasm. To achieve this, a subpopulation of this tRNA is imported into the mitochondrion while still bearing a CCA anticodon. Following import, RNA editing of C$_{34}$ creates the U$_{34}$CA anticodon required to translate the mitochondrial UGA codons as tryptophan (Alfonzo et al. 1999; Simpson et al. 2000; Fig. 4). Despite the existence of other tRNAs that contain C34, only tRNA^{Trp} undergoes editing, raising the question as to what determines specificity. Mass spectrometry analysis of native cytosolic and mitochondrial versions of tRNA^{Trp} revealed that this tRNA undergoes a number of mitochondria-specific modifications following mitochondrial import (Crain et al. 2002). We have raised the possibility that these modifications occur in a cascade, and only when they occur in the proper sequence can editing be specified. In this proposal, although other tRNAs may contain the same modification set, only tRNAs in which modifications occur in the correct sequence, combined with other determinants present only in tRNA^{Trp}, serve in part as the basis of editing specificity. Indeed, in this system, we have identified a number of key determinants for tRNA editing, including unique nucleotides at the anticodon loop as well as single base pairs at the acceptor stem. Nevertheless, the editing enzyme still remains elusive, and neither in the marsupial system described above nor in the trypanosomatid system has a functional assay been established for in vitro editing, precluding identification of this important enzyme.

In the examples listed above, it is clear that a prevailing mechanism for reassigning the identity of a tRNA involves nucleotide transitions at the anticodon nucleotides. However, the chemical mechanism by which C to U transitions occur in the anticodon of tRNAs is to date not clear, as the enzyme(s) performing this type of editing has (have) remained elusive. Much work is still needed to identify these activities and clarify the mechanism. Also as seen in the case of C to U editing, although many other tRNAs contain encoded cytosines in the anticodon, the editing enzyme is able to discriminate and edit only the specific tRNA. Future work should also help elucidate the basis of specificity and discrimination of different substrates ensuring fidelity of the process.

4 Lysidine Formation in tRNA: How an Old Modification Became a New Type of Editing

Following the elucidation of the genetic code and proposal of the wobble hypothesis, scientists naturally proceeded to test the various assumptions suggested by the newly established code and the proposed pairing rules. Early studies reported the noticeable absence of a tRNA[Ile] that could decode the AUA isoleucine codons in *E. coli* (Takemura et al. 1969). The only tRNA[Ile] sequenced at the time contained an unmodified G at position 34 (the wobble position). This tRNA could recognize both the U- and C-ending codons for isoleucine, but could not decode the AUA codons. However, it had been demonstrated, using an *E. coli* in vitro translation system, that the codon AUA indeed specified isoleucine (Gardner et al. 1962). Nishimura and co-workers reasoned that if the AUA codons are rare, then maybe they are decoded by a minor tRNA[Ile] species. This led to the purification of a minor tRNA from *E. coli* with an unusual, modified nucleotide at position 34 (the wobble base), which they called N+ (Harada and Nishimura 1974). Although these researchers demonstrated the ability of this minor species to support specific aminoacylation with isoleucine, the chemical nature of the modification was not immediately elucidated.

At about the same time, sequences revealed the presence of a tRNA[Ile] specific for the AUA codon (Scherberg and Weiss 1972). Surprisingly, the only tRNA in this genome that could recognize this codon had a CAU anticodon, which according to the coding rules should decode the canonical AUG codons for methionine (similar findings were also reported in chloroplasts). Yokoyama and co-workers, in collaboration with Nishimura, then used a combination of NMR and mass spectrometry to elucidate the chemical structure of the N+ nucleotide, a cytosine derivative that could account for the apparently aberrant decoding. They found that

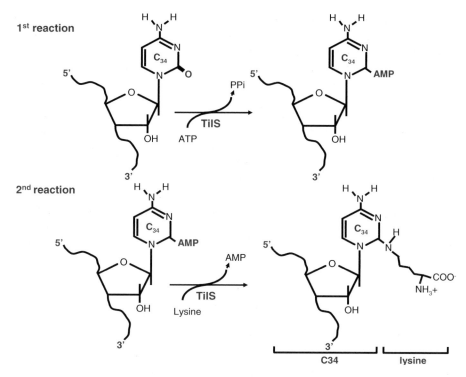

Fig. 5 Proposed two-step mechanism for lysidine formation. *TilS* Lysidine synthetase responsible for catalyzing both steps of the reaction

this new nucleoside contained an unusual lysine side chain at the C2 position of the pyrimidine ring, and termed it "lysidine" (Muramatsu et al. 1988b; Fig. 5).

Functional studies revealed that the native lysidine-containing tRNA[Ile], despite the presence of a CAU anticodon, could be efficiently aminoacylated by isoleucine. In fact, this tRNA could not support charging with methionine, as its anticodon sequence might have suggested. In addition, replacement of lysidine for cytidine, in an otherwise native tRNA, led to a strong reduction in isoleucine accepting activity, and an increase in methionine accepting activity. Combined, these observations led Yokoyama and co-workers to conclude that the single change of cytidine to lysidine could lead to the simultaneous conversion of both the codon and amino acid specificity of tRNA[Ile] (Muramatsu et al. 1988a).

Although not proven at a chemical level, the presence of a lysidine-like change in the anticodon of tRNA[Ile] (CAU) has been suggested to occur in a number of other systems, including spinach chloroplast and bean mitochondria (Weber et al. 1990). However, it is only in potato mitochondria that the presence of a yet unidentified C34 modification also changes the identity of the tRNA[Ile] (CAU) in a manner similar to that described for the bacterial system (Weber et al. 1990). In addition, 13 different archaeal genomes contain tRNA[Ile] with a CAU anticodon, suggesting the possible use of lysidine (Sprinzl and Vassilenko 2005). However, mass spectrometry analysis of total tRNA preparations from Archaea revealed the absence of a species with a molecular mass corresponding to that of lysidine. Therefore, in these organisms, a different lysidine-like modification may exist for the reassignment of the CAU anticodon from Met to Ile.

A major breakthrough in the study of lysidine biosynthesis comes from a recent report from Suzuki and co-workers who, using a bioinformatics approach, identified the gene encoding lysidine synthetase in bacteria (Soma et al. 2003). By searching the Cluster of Orthologous Groups (COGs) database, they identified 48 predictably essential genes in *E. coli*, which at the time had no assigned function. Screening available conditional mutants for a number of these potential genes led to the identification of one locus responsible for lysidine formation (the yacA gene). In the absence of yacA expression, lysidine disappeared from tRNA^{Ile}, as determined by the lack of an activity for the direct incorporation of lysine into tRNA^{Ile} in the conditional mutants, and confirmed by mass spectrometry (Soma et al. 2003; Ikeuchi et al. 2005; Nakanishi et al. 2005).

The product of the yacA gene expressed recombinantly (lysidine synthetase, TilS) was by itself sufficient to synthesize lysidine in an ATP-dependent manner, requiring only in vitro transcribed tRNA^{Ile} and lysine (Soma et al. 2003). These studies have led to a proposed two-step reaction mechanism by which TilS initially catalyzes the activation of the C-2 position of C34 in tRNA^{Ile}, followed by a nucleophilic attack by the ε-amino group of the lysine side chain (Nakanishi et al. 2005; Fig. 5). Interestingly, while lysidine formation involves the incorporation of an amino acid into the anticodon of tRNA^{Ile}, the proposed reaction mechanism does not follow that of aminoacyl-tRNA synthetases. Rather, these enzymes activate the amino acid by formation of an aminoacyl-adenylate, followed by subsequent ligation of the amino acid to the tRNA substrate with the release of AMP. It has been suggested that TilS owes its lysidine synthetase activity to its evolutionary path. The TilS sequence shares great similarity with a family of P-loop PPi synthetases (e.g., GMP synthetase) that analogously activate a nucleoside substrate by adenylation, followed by displacement of the adenylate by a secondary nucleophilic attack from the amino group of lysine (Nakanishi et al. 2005).

In terms of substrate discrimination, it is of interest to note that the bacterial tRNA^{Ile} contains a full set of positive determinants required for methionine incorporation by Met-RS, including the base pairs G2-C71, C3-G70, A73, and the CAU anticodon in addition to key determinants for Ile-RS. Recent results further confirmed that lysidine formation at C34 serves not only to create a strong determinant for Ile-RS recognition, but also as a negative determinant for Met-RS (Ikeuchi et al. 2005). In this manner, the C34 to k^{2}C34 editing event is essential for converting a tRNA^{Ile}, with an encoded Met anticodon, to specify isoleucine by changing its amino acid acceptance. The presence of lysidine also affects the tRNA's decoding capacity, while preventing the undesirable formation of a Met-tRNA^{Ile} product that could have deleterious consequences in translation.

Curiously, the story of lysidine, a study of an unusual modification in a minor tRNA of *E. coli* that began as an attempt to explain the newly elucidated genetic code, now ends as a new type of tRNA editing. In terms of evolution, one can easily infer that for lysidine formation a conserved chemical mechanism utilized by nucleotide biosynthetic enzymes has been recruited by the newly discovered lysidine synthetase to provide the reassignment of a tRNA, with an unusual anticodon, in a process that is essential for cell viability.

5   When Editing Affects Structure: the Use of Editing to Reshape and Repair tRNAs

The few editing mechanisms described above directly affect tRNA function by creating recognition elements for aminoacyl tRNA synthetases, leading to motifs for further modification and/or affecting decoding. However, the majority of editing mechanisms reported to date do not directly affect tRNA identity, but restore tRNA function. These events encompass a number of different mechanisms that include single nucleotide transitions and transversions as well as the replacement of multiple nucleotides. Mechanistically, this group includes nucleotide insertions by nucleotidyltransferase-like activities, polyadenylation, and RdRp-like activities at the 5′ and 3′ ends of tRNAs, as well as a number of yet to be identified activities that act internally in the case of single nucleotide substitutions (Fig. 6).

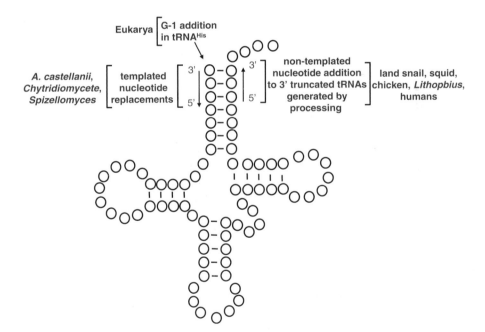

Fig. 6 Structural editing of tRNAs in various systems. Schematic summarizing the different types of editing that serve to repair or restore the structure of tRNAs

The first example of tRNA editing ever reported is that of the amoeboid protozoan *Acanthamoeba castellanii* (Lonergan and Gray 1993). Lonergan and Gray showed that certain tRNAs in the mitochondria have sequences that differ from those found in their encoding genes. These changes thus occur posttranscriptionally, and help restore single or multiple mismatches in the acceptor stem of tRNAs in this organism. Unlike the single nucleotide changes described above involving C to U or A to I changes, in *A. castellanii* editing converts U to A, U to G, and A to G. Since some of these changes involve nucleotide transversions, mechanistically they cannot occur by a simple deamination, as proposed for marsupial and trypanosomatid C to U editing and demonstrated for inosine formation. The observation of multiple nucleotide mismatches in a contiguous stretch of sequence led Gray and co-workers to propose a nucleotide addition mechanism whereby the mismatched nucleotides are removed from the 5′ end of the tRNA, and replaced by nucleotides that recreate canonical base pairs (Price and Gray 1999; Bullerwell and Gray 2005). This mode of nucleotide insertion is, of course, unusual in that it requires the 3′ to 5′ templated addition of nucleotides, where the sequences on the 3′ side of the acceptor stem will provide the editing information. To date, the only example of a non-templated addition of a nucleotide in the 3′ to 5′ direction is that of the non-templated addition of guanosine in tRNA[His] (Fig. 6).

With their unorthodox nucleotide addition mechanism in mind, the Gray laboratory set out to characterize this novel editing activity. They found that *A. castellanii* mitochondrial extracts could indeed edit a synthetic tRNA (Price and Gray 1999). Most recently, in a report of a similar type of editing in the Chytridiomycete fungus, *Spizellomyces punctatus*, these researchers demonstrated the presence of a similar activity (Bullerwell and Gray 2005). Both activities require ATP or a 5′ triphosphate to allow nucleotide incorporation. Mechanistically, it was proposed that the 5′ nucleotide to which an incoming nucleotide is to be added has to be first activated by AMP. This adenylated species is subsequently attacked by the incoming nucleotide, with a concomitant release of AMP. This mechanism implies the need for an activity that can remove the mismatched nucleotides prior to nucleotide addition. Although removal of radioactively labeled 5′ mismatched nucleotides has not been conclusively demonstrated, a 5′ to 3′ exonuclease activity is a likely candidate for such an activity.

In a number of systems, either tRNAs with 3′ end mismatched nucleotides are generated during synthesis, or truncated tRNAs are created during processing in mitochondria. Generally, these tRNAs are repaired by the non-templated addition of adenosine, presumably by poly A polymerase, as is the case for tRNA repair in snail, squid, and chicken mitochondria (Yokobori and Pääbo 1995a, 1995b). In animal mitochondria, tRNA genes are often encoded as overlapping cassettes where the processing of the downstream tRNA yields an upstream tRNA with a shortened 3′ end. Mörl and co-workers showed that these tRNAs are repaired by a CCA-nucleotidyl transferase-like activity, but still in a non-templated manner (Reichert et al. 1998; Reichert and Mörl 2000).

Recently, genes that encode tRNAs with mismatches at the 3′ end of the acceptor stems have been reported in the mitochondria of the centipede *Lithobius forficatus*

(Lavrov et al. 2000), and of *Seculamonas ecuadoriensis* (a jakobid; Leigh and Lang 2004). In these systems, mismatched nucleotides are replaced by nucleotides that regenerate canonical base pairs. However, neither the activity nor an in vitro assay have to date been established. It has been proposed that due to the templated nature of this 3′ addition, an RNA-dependent RNA polymerase could represent a perfect candidate enzyme for this type of editing.

Besides editing events that insert or polymerize multiple nucleotides at tRNA ends or at internal sites, editing by changing single nucleotides to regenerate important structural features has also been known for many years. This includes C to U substitution at the acceptor stems of tRNAs from various plant species. In addition, Marechal-Drouard and co-workers have discovered a single C to U change at position 28 of plant mitochondrial tRNAs (Fey et al. 2002). Curiously, this editing event occurs at a site in the anticodon stem where pseudouridine (Ψ) is found in the mature tRNA. This discovery has led to the proposal of a two-step model for the formation of Ψ, where C to U editing occurs first, presumably by a deamination mechanism, followed by isomerization of the U to form Ψ. Another interesting aspect of plant mitochondrial tRNA editing is the observation that tRNA end-trimming and editing are sequential events (Marechal-Drouard et al. 1996; Kunzmann et al. 1998). Editing of a mismatched nucleotide at the 4th position of the acceptor stem follows 5′ trimming, but precedes 3′ maturation by ribonuclease Z. In this case, editing not only restores structures that might be important for synthetase recognition, but may also serve as a checkpoint for complete maturation of tRNAs.

Most recently, the Marechal-Drouard group tested the fate of the unedited tRNA in potato mitochondria. They introduced, via direct uptake, a DNA template encoding a non-edited version of larch tRNA[His]. They found not only that the potato mitochondria failed to edit this "foreign" tRNA, but also that the unedited tRNA was quickly degraded (Placido et al. 2005). This again emphasizes the possibility that tRNA editing might play a role in ensuring quality control during processing. Evolutionarily, the idea that the potato editing machinery cannot recognize a foreign tRNA could also pose a hurdle for the horizontal transfer of tRNA genes from different plant species.

Mechanisms that repair or restore the structure of tRNAs represent by far the most common and varied types of tRNA editing. As in the case of single nucleotide substitutions, however, in most cases of tRNA repair either the enzyme has not been identified, or an in vitro editing system does not exist. Thus, one can only speculate on the mechanisms that operate in these types of editing events.

6   Concluding Remarks

Transfer RNAs undergo by far the largest number of mechanistically different editing events, compared to other RNAs within a cell. As summarized here, these include nucleotide additions, deaminations, and aminoacylation. It is not at all surprising that throughout evolution, cells have adopted various means of ensuring the availability of functional tRNAs, given the central role played by this molecule in cellular metabolism. Based on this chapter, it is clear that much progress has been made in the identification of some of the tRNA editing enzymes, e.g., in the case of the A to I and the lysidine biosynthetic enzymes. There remains, however, a long road to be traveled. For instance, the C to U editing enzymes are largely still unknown, and since currently no in vitro assay exists for this type of editing, one can only speculate about a possible mechanism. Efforts to identify the 5′ and 3′ nucleotide adding enzymes have progressed further and very efficient in vitro assays are now available, which should lead to their more detailed characterization. However, identification of editing activities is only the proverbial "tip of the iceberg", as much biochemistry is still needed to define how these enzymes recognize their substrates. In the case of tRNA, the question of RNA binding and substrate recognition is of particular importance, seeing that tRNA editing enzymes often face the problem of specific substrate discrimination in a pool of very similar (nearly identical) substrates. This conundrum becomes especially apparent in the case of single nucleotide substitutions where the editing enzyme likely relies on minor changes in local substrate structure to achieve a high degree of specificity.

Despite the many shortcomings, the tRNA editing field is moving fast, and with the advent of the genomics and bioinformatics era, there is no doubt that many important questions will soon be answered in this domain. This chapter has gathered and discussed a number of tRNA editing examples. This list is by no means conclusive or exhaustive, and so, rather than closing the chapter on tRNA editing, this review has attempted to open a window onto different approaches that have led to the discovery of new editing types. It is my hope that this serve as a source of motivation and a call to arms for new efforts toward the discovery of many more examples of tRNA editing, as well as the elucidation of the multifaceted associated mechanisms.

References

Alfonzo JD, Blanc V, Estevez AM, Rubio MA, Simpson L (1999) C to U editing of the anticodon of imported mitochondrial tRNA(Trp) allows decoding of the UGA stop codon in *Leishmania tarentolae*. EMBO J 18:7056–7062

Auxilien S, Crain PF, Trewyn RW, Grosjean H (1996) Mechanism, specificity and general properties of the yeast enzyme catalysing the formation of inosine 34 in the anticodon of transfer RNA. J Mol Biol 262:437–458

Benne R, Van den Burg J, Brakenhoff JP, Sloof P, Van Boom JH, Tromp MC (1986) Major transcript of the frameshifted coxII gene from trypanosome mitochondria contains four nucleotides that are not encoded in the DNA. Cell 46:819–826

Börner GV, Mörl M, Janke A, Pääbo S (1996) RNA editing changes the identity of a mitochondrial tRNA in marsupials. EMBO J 15:5949–5957

Bullerwell CE, Gray MW (2005) In vitro characterization of a tRNA editing activity in the mitochondria of *Spizellomyces punctatus*, a Chytridiomycete fungus. J Biol Chem 280:2463–2470

Covello PS, Gray MW (1998) Editing of tRNA. In: Grosjean H, Benne R (eds) Modification and editing of tRNA. ASM Press, Washington, DC, pp 596

Crain PF, Alfonzo JD, Rozenski J, Kapushoc ST, McCloskey JA, Simpson L (2002) Modification of the universally unmodified uridine-33 in a mitochondria-imported edited tRNA and the role of the anticodon arm structure on editing efficiency. RNA 8:752–761

Crick FH (1966) Codon–anticodon pairing: the wobble hypothesis. J Mol Biol 19:548–555

Fey J, Weil JH, Tomita K, Cosset A, Dietrich A, Small I, Marechal-Drouard L (2002) Role of editing in plant mitochondrial transfer RNAs. Gene 286:21–24

Gardner RS, Wahba AJ, Basilio C, Miller RS, Lengyel P, Speyer JF (1962) Synthetic polynucleotides and the amino acid code. VII Proc Natl Acad Sci USA 48:2087–2094

Gerber AP, Keller W (1999) An adenosine deaminase that generates inosine at the wobble position of tRNAs. Science 286:1146–1149

Grosjean H, Björk GR (2004) Enzymatic conversion of cytidine to lysidine in anticodon of bacterial isoleucyl-tRNA – an alternative way of RNA editing. Trends Biochem Sci 29:165–168

Grosjean H, Auxilien S, Constantinesco F, Simon C, Corda Y, Becker HF, Foiret D, Morin A, Jin YX, Fournier M, Fourrey JL (1996a) Enzymatic conversion of adenosine to inosine and to N1-methylinosine in transfer RNAs: a review. Biochimie 78:488–501

Grosjean H, Edqvist J, Straby KB, Giege R (1996b) Enzymatic formation of modified nucleosides in tRNA: dependence on tRNA architecture. J Mol Biol 255:67–85

Harada F, Nishimura S (1974) Purification and characterization of AUA specific isoleucine transfer ribonucleic acid from Escherichia coli B. Biochemistry 13:300–307

Holley RW, Apgar J, Everett GA, Madison JT, Marquisee M, Merrill SH, Penswick JR, Zamir A (1965a) Structure of a ribonucleic acid. Science 147:1462–1465

Holley RW, Everett GA, Madison JT, Zamir A (1965b) Nucleotide sequences in the yeast alanine transfer ribonucleic acid. J Biol Chem 240:2122–2128

Ikeuchi Y, Soma A, Ote T, Kato J, Sekine Y, Suzuki T (2005) Molecular mechanism of lysidine synthesis that determines tRNA identity and codon recognition. Mol Cell 19:235–246

Janke A, Pääbo S (1993) Editing of a tRNA anticodon in marsupial mitochondria changes its codon recognition. Nucleic Acids Res 21:1523–1525

Janke A, Feldmaier-Fuchs G, Thomas WK, von Haeseler A, Pääbo S (1994) The marsupial mitochondrial genome and the evolution of placental mammals. Genetics 137:243–256

Kunzmann A, Brennicke A, Marchfelder A (1998) 5' end maturation and RNA editing have to precede tRNA 3' processing in plant mitochondria. Proc Natl Acad Sci USA 95:108–113

Kuratani M, Ishii R, Bessho Y, Fukunaga R, Sengoku T, Shirouzu M, Sekine S, Yokoyama S (2005) Crystal structure of tRNA adenosine deaminase (TadA) from Aquifex aeolicus. J Biol Chem 280:16002–16008

Lavrov DV, Brown WM, Boore JL (2000) A novel type of RNA editing occurs in the mitochondrial tRNAs of the centipede Lithobius forficatus. Proc Natl Acad Sci USA 97:13738–13742

Leigh J, Lang BF (2004) Mitochondrial 3' tRNA editing in the jakobid Seculamonas ecuadoriensis: a novel mechanism and implications for tRNA processing. RNA 10:615–621

Lonergan KM, Gray MW (1993) Editing of transfer RNAs in Acanthamoeba castellanii mitochondria. Science 259:812–816

Marechal-Drouard L, Cosset A, Remacle C, Ramamonjisoa D, Dietrich A (1996) A single editing event is a prerequisite for efficient processing of potato mitochondrial phenylalanine tRNA. Mol Cell Biol 16:3504–3510

Mörl M, Dorner M, Pääbo S (1995) C to U editing and modifications during the maturation of the mitochondrial tRNA(Asp) in marsupials. Nucleic Acids Res 23:3380–3384

Muramatsu T, Nishikawa K, Nemoto F, Kuchino Y, Nishimura S, Miyazawa T, Yokoyama S (1988a) Codon and amino-acid specificities of a transfer RNA are both converted by a single post-transcriptional modification. Nature 336:179–181

Muramatsu T, Yokoyama S, Horie N, Matsuda A, Ueda T, Yamaizumi Z, Kuchino Y, Nishimura S, Miyazawa T (1988b) A novel lysine-substituted nucleoside in the first position of the anticodon of minor isoleucine tRNA from Escherichia coli. J Biol Chem 263:9261–9267

Nakanishi K, Fukai S, Ikeuchi Y, Soma A, Sekine Y, Suzuki T, Nureki O (2005) Structural basis for lysidine formation by ATP pyrophosphatase accompanied by a lysine-specific loop and a tRNA-recognition domain. Proc Natl Acad Sci USA 102:7487–7492

Placido A, Gagliardi D, Gallerani R, Grienenberger JM, Marechal-Drouard L (2005) Fate of a larch unedited tRNA precursor expressed in potato mitochondria. J Biol Chem (in press)

Price DH, Gray MW (1999) A novel nucleotide incorporation activity implicated in the editing of mitochondrial transfer RNAs in *Acanthamoeba castellanii*. RNA 5:302–317

Reichert AS, Mörl M (2000) Repair of tRNAs in metazoan mitochondria. Nucleic Acids Res 28:2043–2048

Reichert A, Rothbauer U, Mörl M (1998) Processing and editing of overlapping tRNAs in human mitochondria. J Biol Chem 273:31977–31984

Scherberg NH, Weiss SB (1972) T4 transfer RNAs: codon recognition and translational properties. Proc Natl Acad Sci USA 69:1114–1118

Simpson L, Thiemann OH, Savill NJ, Alfonzo JD, Maslov DA (2000) Evolution of RNA editing in trypanosome mitochondria. Proc Natl Acad Sci USA 97:6986–6993

Soma A, Ikeuchi Y, Kanemasa S, Kobayashi K, Ogasawara N, Ote T, Kato J, Watanabe K, Sekine Y, Suzuki T (2003) An RNA-modifying enzyme that governs both the codon and amino acid specificities of isoleucine tRNA. Mol Cell 12:689–698

Sprinzl M, Vassilenko KS (2005) Compilation of tRNA sequences and sequences of tRNA genes. Nucleic Acids Res 33 Database Issue:D139–140

Takemura S, Murakami M, Miyazaki M (1969) Nucleotide sequence of isoleucine transfer RNA from *Torulopsis utilis*. J Biochem (Tokyo) 65:489–491

Urbonavicius J, Qian Q, Durand JM, Hagervall TG, Björk GR (2001) Improvement of reading frame maintenance is a common function for several tRNA modifications. EMBO J 20:4863–4873

Urbonavicius J, Stahl G, Durand JM, Ben Salem SN, Qian Q, Farabaugh PJ, Björk GR (2003) Transfer RNA modifications that alter +1 frameshifting in general fail to affect −1 frameshifting. RNA 9:760–768

Weber F, Dietrich A, Weil JH, Marechal-Drouard L (1990) A potato mitochondrial isoleucine tRNA is coded for by a mitochondrial gene possessing a methionine anticodon. Nucleic Acids Res 18:5027–5030

Wolf J, Gerber AP, Keller W (2002) tadA, an essential tRNA-specific adenosine deaminase from *Escherichia coli*. EMBO J 21:3841–3851

Yokobori S, Pääbo S (1995a) Transfer RNA editing in land snail mitochondria. Proc Natl Acad Sci USA 92:10432–10435

Yokobori SI, Pääbo S (1995b) tRNA editing in metazoans. Nature 377:490

RNA Editing by Adenosine Deaminases that Act on RNA (ADARs)

Michael F. Jantsch[1] (✉) and Marie Öhman[2]

[1] Department of Chromosome Biology, Max F. Perutz Laboratories, University of Vienna, Dr. Bohr Gasse 1, 1030 Vienna, Austria; Michael.Jantsch@univie.ac.at
[2] Department of Molecular Biology & Functional Genomics, Stockholm University, S-106 91 Stockholm, Sweden; Marie.Ohman@molbio.su.se

H.U. Göringer (ed.), *RNA Editing. Nucleic Acids and Molecular Biology 20*
© Springer-Verlag Berlin Heidelberg 2008

> *"The moment I saw the model and heard about the comple-*
> *menting base pairs I realized that it was the key to understand-*
> *ing all the problems in biology we had found intractable – it*
> *was the birth of molecular biology."*
>
>   *Sydney Brenner*

Abstract Adenosine deaminases that act on RNA (ADARs) give rise to the most abundant form of RNA editing found in Metazoa. ADAR proteins convert adenosines to inosines within structured and double-stranded RNAs. Since inosines are interpreted as guanosines by several cellular machineries, the consequences of editing can be widespread. In messenger RNA, alterations of codons, changes in splice patterns, and influences on RNA stability have been observed as a result of RNA editing. Moreover, A to I editing has been shown to interconnect with the RNA interference machinery. In this chapter, an overview on ADAR enzymes, their molecular architecture, occurrence, and substrate specificity is given. Consequences of editing, studies in model organisms, and implications for other double-stranded RNA-dependent processes are discussed.

1   Introduction

Adenosine deaminases that act on RNA (ADARs) convert adenosines to inosines via hydrolytic deamination at the C6 position of adenosines. As adenosines are interpreted as guanosines during translation, this nucleotide change can lead to a codon exchange in the coding region of the affected RNA (Bass 2002; Fig. 1).

ADAR-mediated editing is a widespread phenomenon that appears to exist in all metazoans, and is important for normal life and development. Animals in which ADARs have been deleted show a broad spectrum of phenotypes; invertebrates lacking ADARs are viable but exhibit behavioral defects (Palladino et al. 2000a; Tonkin et al. 2002), while lack of ADARs in vertebrates can lead to embryonic lethality or severe neurological pathologies (Higuchi et al. 2000; Hartner et al. 2004; Wang et al. 2004).

ADARs bind their substrates via double-stranded RNA-binding domains (dsRBDs) that are harbored by all members of this protein family, and consistently, ADAR targets are defined by structured or double-stranded regions present in edited RNAs. The extent of editing of a particular RNA molecule depends largely on the length and structure of the double-stranded region (Lehmann and Bass 1999). Short stem-loop structures frequently define regions where only single nucleotides are edited, while longer, more extensive double-stranded structures often contain multiple editing sites, leading to a phenomenon referred to as hyperediting.

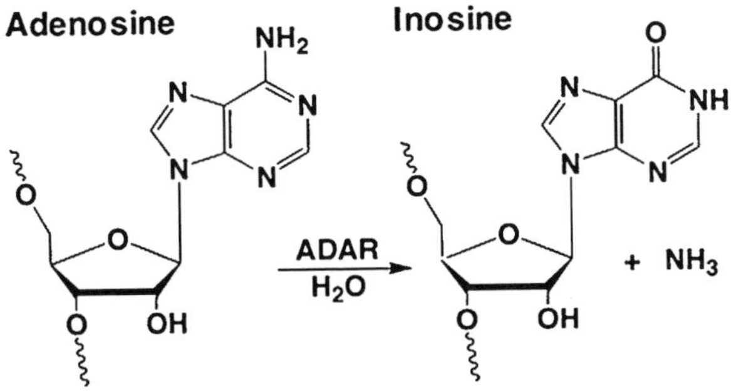

Fig. 1 ADARs catalyze hydrolytic deamination at the C6 position of adenosine. The resulting product, inosine, base-pairs with cytosine and thus has the coding potential of a guanosine residue. (Reprinted with permission from Macbeth et al. 2005, Science 309:1534–1539. Copyright 2005, AAAS)

Single-site editing occurs mostly in coding regions of particular mRNAs, while hyperediting frequently affects viral RNAs but also untranslated regions of mRNAs.

Substrate specificity is mediated by the catalytic domain. The recently solved crystal structure of this domain provides interesting insights into substrate binding, and the catalysis of the reaction (Macbeth et al. 2005).

The frequently observed proximity of editing sites to splice sites, and the fact that intronic sequences are sometimes required to define editing sites indicate a crosstalk between splicing and editing. In fact, recent experimental data provide evidence for coordination of both processes.

Although biological and biochemical evidence suggests a rather widespread occurrence of editing, until recently only a few editing targets were known. Indeed, novel bioinformatics data detected widespread editing in transcribed repetitive elements, the function of which remains enigmatic at this point (Nishikura 2004). In addition, microRNAs have been identified as targets for ADAR-mediated RNA editing, and similarly a crosstalk between RNA editing and the RNAi machinery has been demonstrated (Luciano et al. 2004; Yang et al. 2005; Blow et al. 2006; Kawahara et al. 2007). Thus, A to I RNA editing seems to have a wide repertoire of biological functions ranging from the alteration of individual protein sequences, over microRNA-mediated gene regulation to epigenetic phenomena such as heterochromatin formation (Fernandez et al. 2005).

2 The Enzymes

The enzymatic activity leading to adenosine deamination was originally described in *Xenopus* oocytes as an RNA unwinding activity (Bass and Weintraub 1988). Meanwhile, this enzymatic activity has been detected in all metazoan tissues tested.

Furthermore, molecular cloning, and the availability of a growing list of fully sequenced genomes have demonstrated the existence of a family of related, yet distinct ADAR proteins having different numbers of family members in different species. The nomenclature for the group of enzymes now commonly referred to as ADARs was originally as heterogeneous as the members of this enzyme family. DRADA, dsRAD, or RED were some of the different names given to the enzymes performing the hydrolytic deamination of adenosines in RNAs (Bass et al. 1997). ADARs are functionally and structurally related to adenosine deaminases that act on tRNA substrates (ADATs; Gerber et al. 1998).

In mammals, three ADAR proteins, ADAR1, ADAR2 and ADAR3, have been isolated, while other vertebrates as well as invertebrates such as *Caenorhabditis elegans* harbor two ADAR variants in their genomes (Tonkin et al. 2002). Only a single ADAR protein exists in *Drosophila* (Palladino et al. 2000b). The different ADAR proteins differ not only with respect to their expression patterns but also in their substrate specificity, and hence in their biological function. Vertebrate ADAR1, for instance, is more ubiquitously expressed than ADAR2, which seems to be restricted to the nervous system (Kim et al. 1994; Melcher et al. 1996b). Consistently, the substrates and biological functions of the different ADAR proteins vary considerably (see below).

3 Molecular Architecture

ADAR proteins are defined by a conserved deaminase domain located at the C-terminal end of the protein that is required for catalysis. A variable number of dsRBDs are found upstream of the catalytic domain. These are responsible for RNA binding, and also contribute to substrate specificity. The N-terminal ends of ADARs are variable; ADAR2 has a very short amino terminus, starting immediately upstream of the dsRBDs, while the amino terminus of ADAR1 is long, harboring Z-DNA-binding domains (ZBDs) and a nuclear export signal (NES). In contrast, the amino terminus of ADAR3 contains a single-stranded R-type RNA-binding domain (Fig. 2).

3.1 The Deaminase Domain and Catalysis

The deaminase domain is related to other nucleotide deaminases, which has led to the suggestion that zinc might be responsible for the coordination of water as a nucleophile for the catalytic reaction (Kim et al. 1994). Similarity to methyltransferases has suggested a mode of action by which the adenosine to be edited is flipped out of its double-stranded context (Hough and Bass 1997). The use of nucleotide analogues has allowed a more detailed analysis of transition-state intermediates of the catalytic reaction. An RNA in which the substrate adenosine was replaced by

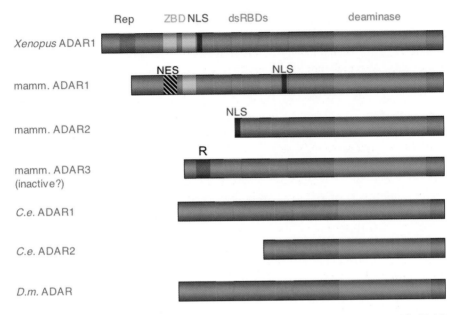

Fig. 2 Schematic organization of ADAR proteins in invertebrates and vertebrates. All ADARs contain a conserved deaminase domain at their C-terminal ends (deaminase, *green*), and a varying number of double-stranded RNA-binding domains (dsRBDs, *red*). The amino termini of ADARs are highly variable. Vertebrate ADAR1 contains Z-DNA-binding domains (ZBD, *yellow*), while mammalian ADAR3 harbors a single-stranded RNA-binding R-motif (R, *magenta*). In addition, nuclear export (NES) and nuclear import signals (NLS) have been defined in several regions of various ADAR proteins. See text for further details

2-aminopurine showed a marked increase in fluorescence upon binding of ADAR2, consistent with the adenosine to be edited being flipped out of its double-stranded context during the editing reaction (Stephens et al. 2000). These findings are also supported by footprinting experiments in which the strand opposing the adenosine to be edited became accessible to hydroxyradical modification based on ADAR2 binding (Yi-Brunozzi et al. 2001).

The deaminase domain of ADAR2 has recently been crystallized, and its structure solved at 1.7 Å resolution, providing new insight into target recognition and the mode of action of this domain. In the crystal, two domains form an asymmetric unit. Nevertheless, these domains behave as monomers, with no evidence for dimer formation in the catalytic part of the enzyme (Macbeth et al. 2005).

As for cytidine deaminases and the tRNA-editing enzyme TadA, a zinc-coordinating deaminase core could be identified. However, while the zinc-coordinating residues are in close proximity on the surface of an alpha helix in TadA and CDA, the equivalent cysteine residues are separated by a 64-amino acid-long loop in ADAR2. Further stabilization of this region is thus required in the catalytic domain of ADARs, which seems to be achieved by an additional hydrogen bond to a lysine residue that is uniquely conserved in all ADARs (Fig. 3).

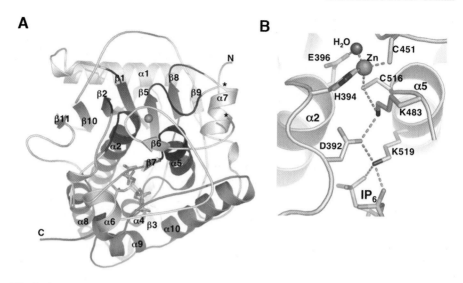

Fig. 3 A Ribbon model of the catalytic deaminase domain of ADAR2 derived from X-ray crystallography data. The active-site zinc atom is shown as a *magenta sphere*. A region sharing structural similarity to cytidine deaminases and ADATs is in *dark blue*. The AC-terminal helical domain, which makes the major contacts to inositol-hexakisphosphate (ball and stick), is in *red*. **B** Residue interactions at the active site. Shown are the zinc ion, coordinating residues (H394, C451, and C516), the nucleophilic water (*blue sphere*), and the proposed proton-shuttling residue, E396. The hydrogen-bond relay that connects the active site to the IP6 is also indicated. (Reprinted with permission from Macbeth et al. 2005, Science 309:1534–1539. Copyright 2005, AAAS)

Interestingly, a unique gap in the structure of the ADAR2 catalytic domain can harbor a molecule of inositol-hexakisphsophate (IP6). The IP6 moiety forms several hydrogen bonds with side chains that are conserved in ADARs and ADAT1 proteins (Fig. 3). These hydrogen bonds also stabilize the conserved lysine near the catalytic core that is involved in coordinating the zinc atom required for the deamination reaction. IP6 co-purifies with the ADAR2 catalytic domain, indicating that it binds tightly to it. In the absence of IP6, the domain presumably misfolds, possibly explaining why ADARs expressed in bacteria are catalytically inactive.

Modeling of the cytidine analog zebularine in the catalytic core of the available ADAR2 structure also nicely explains the specificity of ADARs for adenosines: cytidine binding leads to a steric clash of the ribose with a side chain in the catalytic center. In contrast, the different structure of the adenosine purine ring fits nicely into the catalytic core (Macbeth et al. 2005).

3.2   The Double-Stranded RNA-Binding Domains (dsRBDs)

All ADARs contain one to three double-stranded RNA-binding domains (dsRBDs) upstream of the deaminase domain that are obviously required to bind substrate

RNAs. In vertebrates, ADAR1 has three dsRBDs, while ADAR2 and ADAR3 contain two dsRBDs. In *C. elegans*, on the other hand, Adr-1 has two, while Adr-2 has only one dsRBD (Tonkin et al. 2002). *Drosophila* dADAR also contains two dsRBDs (Palladino et al. 2000b). Besides substrate binding, the dsRBDs also appear to contribute to substrate specificity, and at least in some cases, are essential for protein dimerization (see below).

3.2.1   The Role of dsRBDs in Substrate Recognition

dsRBDs recognize structured or double-stranded RNAs (dsRNAs) approximately 16 base-pairs in length. However, the region bound by a dsRBD can also be interrupted by internal bulges, or consist of shorter stacked helices. For most dsRBDs, contact with the RNA is made at three positions all of which recognize the structure, rather than the actual sequence of the RNA molecule. In contrast to double-stranded DNA (dsDNA), dsRNA has a wide and shallow minor groove while the major groove is narrow and deep, precluding base-specific contacts with amino acid side chains. The contacts between the dsRBD and the RNA molecule are therefore specific for the characteristic distances across the major and minor groove of dsRNA (Ryter and Schultz 1998). Given that no sequence-specific contacts between the dsRBDs and RNA can be made, it was inferred that the dsRBDs contribute only marginally to specific substrate recognition, and that the major part of site specificity derives from the deaminase domain itself. Consistently, domain swapping experiments, in which the deaminase domains of ADAR1 and ADAR2 were exchanged, indicated that some, but not all substrate specificity is determined by the catalytic domain (Wong et al. 2001).

Recent experiments, however, have identified an important function of dsRBDs in substrate recognition and catalysis. For a start, mutational analysis of *Xenopus* ADAR1 has demonstrated the importance of dsRBDs in targeting the protein to transcriptionally active regions. Consistent with its co-transcriptional function, ADAR1 can be found associated with the majority of transcriptionally active sites on *Xenopus* lampbrush chromosomes. Deletion or duplication of individual dsRBDs shifts the distribution of ADAR1 to a subset of transcriptionally active sites, underscoring that individual dsRBDs can specifically target the protein to individual substrates (Doyle and Jantsch 2003). Moreover, binding studies of a short hairpin derived from the Q/R site of glutamate receptor subunit B (gluR-B) indicate selective binding of the two dsRBDs of ADAR2 to this RNA (Stephens et al. 2004). Also, structural analysis of dsRBDs has identified subtle, yet distinct differences between different dsRBDs. The third dsRBD of the *Drosophila* Staufen protein can, for instance, satisfy one of its three RNA contacts by a loop structure (Ramos et al. 2000). Similarly, the RNAse Rnt1p also specifically recognizes its substrate by interacting with a tetraloop structure. Interestingly, the Rnt1p dsRBD structure differs from other dsRBDs by a C-terminal alpha helical extension that slightly alters the overall shape of the dsRBD to facilitate the contact with the tetraloop (Wu et al. 2004). Also the R/G site in the glutamate receptor subunit B pre-mRNA can form a characteristic pentaloop. The specific structure adopted by this loop might also contribute to specific recognition of this substrate (Stefl and Allain 2005). Consistently, the two dsRBDs of ADAR2 have been shown to position each other specifically on a minimal substrate containing the R/G site. While the first dsRBD interacts with the loop structure, the second dsRBD recognizes the stem region containing the adenosine to be edited (Stefl et al. 2006). Some dsRBDs can thus specifically recognize loop regions and asymmetries deviating from perfect dsRNA, and it is likely that ADARs utilize this mechanism to recognize their substrates.

3.3   A Variable Amino Terminus

The largest variability amongst ADARs can be found in their amino terminal ends. While ADAR1 has the longest amino terminus, ADAR2 starts immediately upstream of its first dsRBD, and seems to lack any regulatory elements in its amino terminus. ADAR1 is expressed from two promoters. A constitutive promoter gives rise to a shorter transcript and a protein that starts at Met 296, which thus lacks most of the amino terminus. An interferon-induced version of ADAR1, in contrast, harbors two ZBDs and an NES. In vitro binding studies have shown that the ZBDs can bind to left-handed Z-DNA (Herbert et al. 1997, 1998). It was therefore proposed that the ZBDs facilitate the binding of ADAR1 to transcriptionally active sites where the DNA may be found in left-handed conformation in the immediate proximity of transcription complexes (Kim et al. 2000). However, deletion variants of ADAR1 lacking the ZBDs seem to associate normally with chromosomes and are enzymatically active, thus suggesting that all ZBDs contribute only marginally to chromosomal targeting of the enzyme. Another function of the ZBDs may lie in their contribution to facilitate editing of short substrate RNAs (Herbert and Rich 2001). This finding is of particular interest when considering that microRNA precursors were recently identified as targets for A to I editing by ADAR1 (Luciano et al. 2004; Blow et al. 2006). Consistently, the amino terminus of ADAR1 seems important for the binding of siRNAs, thereby interfering with their efficacy in vivo (Yang et al. 2005).

A classical leucine-rich NES can be found overlapping the first Z-DNA-binding domain at the amino terminus of the interferon-induced, full-length version of ADAR1. This NES is responsible for the continuous nucleo-cytoplasmic shuttling of the enzyme (Poulsen et al. 2001; Strehblow et al. 2002). Nucleo-cytoplasmic shuttling may be required to regulate the nuclear and cytoplasmic concentrations of the enzyme, thereby regulating enzyme activity. ADAR3 harbors a single-stranded RNA-binding R-motif at its amino terminus. This motif might facilitate binding of the enzyme to single-stranded RNA (Chen et al. 2000). However, given the fact that no enzymatic activity has been detected for ADAR3, and that no substrates of this enzyme are known, the functional significance of this motif remains enigmatic at this point.

4 Expression and Occurrence in Different Model Systems

As mentioned above, the activity of A to I adenosine deamination has been found in every metazoan tested, from *C. elegans* to man. In nematodes, the two ADAR genes *adr-1* and *adr-2* are rather uniformly expressed during early embryogenesis, but their expression becomes increasingly restricted with ongoing development. No precise expression data are available for *adr-2*. The expression of *adr-1*, however, is known to be restricted to the nervous system and the developing vulva in late larval stages, and exclusively to nervous tissue in adult worms (Tonkin et al. 2002). The single ADAR gene in *Drosophila* is expressed from two different promoters, and has several different splice variants. Expression of the protein is most prominent in the nervous system (Palladino et al. 2000b), and lack of expression leads to defects primarily in this tissue (Palladino et al. 2000a). Nonetheless, fine-tuning of ADAR activity by alternative splicing and self-editing is essential in *Drosophila* (Keegan et al. 2005).

In mammals, A to I editing activity has also been detected ubiquitously in liver, kidney, spleen, testis, lymph node, and brain (Wagner et al. 1990). ADAR2 is expressed mainly in the brain, although heart, lung, kidney, and testis also express a significant amount of ADAR2 transcript. Although the amounts vary, ADAR1 mRNA is present in all tissues examined to date (O'Connell et al. 1995). However, with few exceptions, most coding substrates that are edited by the two enzymes are in the central nervous system. When the mammalian brain was examined by in situ hybridization for ADAR1 and ADAR2 expression, the expression profile differed between the two enzymes. ADAR1 gene expression is fairly homogeneous throughout the brain. In general, the transcript levels of ADAR2 are higher than for ADAR1, particularly in tissues such as the cerebellum and thalamus. ADAR2 is also expressed in most structures of the brain (Melcher et al. 1996b).

5 The Substrates

5.1 *Glutamate Receptor Pre-mRNA Editing in Mammals*

Of the three glutamate receptor channels (GluRs) NMDA, AMPA and kainate, transcripts of the latter two are edited (reviewed in Bass 2002; Seeburg and Hartner 2003). The best known substrate for ADAR editing is GluR-B of the AMPA receptor. This glutamate receptor is assembled by a heteromeric set of subunits (GluR-A, -B, -C, and -D) creating a cation-selective ion channel, gated by synaptically released glutamate. Most of the fast excitatory neurotransmission in the mammalian brain is mediated by this receptor. Neurons control their level of Ca^{2+} influx by the number of AMPA receptors assembled with the GluR-B subunit. High ratios of GluR-B to the other subunits exhibit low permeability to Ca^{2+} (Geiger et al. 1995).

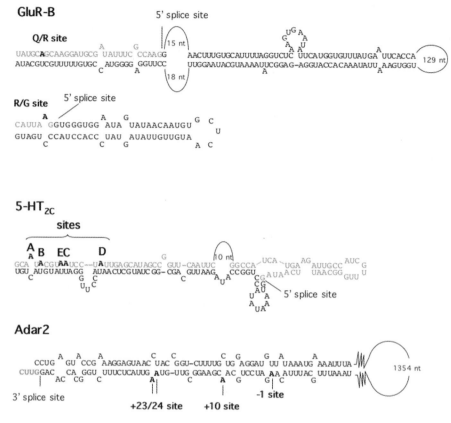

Fig. 4 Predicted RNA secondary structures for a selection of mammalian A to I editing substrates. Site-selectively edited nucleotides are indicated in *bold*. Exon sequence is shown in *grey*, while intron sequence is in *black*. GluR-B is the glutamate receptor subunit B, 5-HT$_{2C}$ the serotonin receptor subtype 2C, and Adar 2 the pre-mRNA expressing the ADAR2 protein (Dawson et al. 2004)

ADARs target the coding sequence of the GluR-B transcript at two positions, the Q/R and the R/G sites (Fig. 4). After A to I editing at the Q/R site, a CAG codon for glutamine (Q) is changed to CIG, which codes for arginine (R). Editing at the Q/R site is essential. Mice that lack all editing at this site suffer from epileptic seizures and die within 3 weeks of birth (Brusa et al. 1995). More than 99% of the GluR-B subunits are in the edited R form. The Q/R site is situated in transmembrane region M2 of the ion channel, critically changing the environment so that the channel becomes impermeable to Ca^{2+} (Sommer et al. 1991). Receptor subunits, which contain glutamine in the homologous position, assemble into channels with a high Ca^{2+} permeability. Interestingly, assembly and subsequent export from the endoplasmatic reticulum is specifically regulated for the different GluR-B isoforms (Greger et al. 2002, 2003).

The Q/R editing site also exists in kainate receptor subunits GluR-5 and -6. However, editing at these sites is at a lower level and not essential, and therefore affects the properties of the receptor to a lesser degree. AMPA receptor subunits

GluR-B, -C, and -D are also edited at the R/G site, with editing levels ranging from 50 to 95% (Seeburg et al. 1998). Receptors assembled by subunits of the R-form (arginine = R; AGA) and the G-form (glycine = G; IGA) have different recovery rates from desensitization (Lomeli et al. 1994).

5.2 Serotonin Receptor Pre-mRNA Editing in Mammals

The neurotransmitter serotonin interacts with several receptor subtypes of the 5-HT$_2$ family. These G-protein coupled receptors are of three subtypes: 5-HT$_{2A}$, 5-HT$_{2B}$, and 5-HT$_{2C}$. Transcripts encoding the 2C subtype undergo A to I editing at five sites: A, B, E, C, and D, situated in close proximity to each other (Fig. 4). Editing at the A site, or at the A and B sites converts an isoleucine into a valine at position 156 of the human receptor. Editing solely at the B site generates a methionine at the same position. Asparagine 158 is changed to serine upon editing at the C site, and editing at both C and E generates a glycine. Finally, editing at the D site results in the substitution of a valine for an isoleucine at position 160. Theoretically, 24 different receptor variants can be produced from different combinations of the edited sites. Indeed, at least 10 different isoforms have been shown to be produced in various amounts in different parts of the rat brain, and 12 isoforms in different human brain regions (Burns et al. 1997; Wang et al. 2000). However, the editing pattern differs considerably between rodents and humans, suggesting that the isoforms specific to the different species may have distinct functional roles (Niswender et al. 1999).

5.3 Potassium Channel mRNA Editing

Potassium channels of the K$_v$1 subfamily are involved in repolarizing membranes, and determining the firing properties of neurons. The transcript of the human K$_v$1.1 gene is subjected to A to I editing at a single position, causing an isoleucine (I) to valine (V) amino acid change in the translated protein (Hoopengardner et al. 2003). The change lies in the ion-conducting pore of the channel. Voltage-dependent potassium channels modulate the excitability by opening and closing the K$^+$-selective pore in response to voltage. A process known as fast inactivation can occur when the ion flow is interrupted by an intracellular particle. Editing at the I/V site of K$_v$1.1 affects the process of fast inactivation (Bhalla et al. 2004). The I to V conversion results in a 20-fold increase in recovery rate from fast inactivation.

Potassium channel editing in humans was first discovered by comparative genomics. Directed by the neurological phenotype of *Drosophila melanogaster* dADAR null mutants and phylogenetically conserved regions among several *Drosophila* species, 914 genes expressed in the nervous system of the fly were analyzed for A to I editing (Hoopengardner et al. 2003). Editing found in the transcript of the *Drosophila* Shaker gene was compared with that of the human, mouse, and rat Shaker genes. An unusual conservation between the mammalian $K_v1.1$ orthologs was found, which led to the discovery of edited sites in these species. Subsequent analysis of RNA editing within this region revealed the A to I edited I/V site.

Transcripts coding for two potassium channel subunits in the squid giant axon have been found to be extensively A to I edited. In the K_v2 K⁺-channel, membrane segments S4–S6 are subjected to editing at 17 sites (Patton et al. 1997). Two of these sites influence the rate of closing and inactivation of the channel. The $SqK_v1.1A$ mRNA was found to be edited at 14 sites (Rosenthal and Bezanilla 2002). Half of the sites are situated in the T1 domain, important for receptor assembly, and the other sites are in transmembrane regions, with the potential to affect channel gating. Another interesting hypothesis of Rosenthal and Bezanilla (2002) is that some of these edited sites in squid K⁺-channels, but also in other proteins, arise from temperature adaptation. Since the amino acid changes caused by editing resemble mutations arising from cold-temperature adaptation in other proteins, it is possible that the squid also uses editing to adapt to changes in ambient seawater temperature.

5.4 Editing as an Auto-Regulatory Mechanism

A functional A to G change in the pre-mRNA can have consequences other than only codon change. One modification that has been postulated to regulate the expression of ADAR2 is the auto-editing of the ADAR2 pre-mRNA (Fig. 4). Since I is read as a G by the splicing machinery, editing has the potential to regulate splicing at several locations. In the ADAR2 pre-mRNA, an A to I modification within intron 4 changes an AA dinucleotide into an AI dinucleotide, which gives rise to an alternative 3′ splice site (Rueter et al. 1999). The result of this modification is alternative splicing. Since the alternative 3′ splice site is situated 47 nucleotides upstream of the normal splice site, within intron 4, alternative splicing will lead to a frame shift upon translation. Frequently, the frame shift leads to subsequent termination of translation at premature stop codons. The alternative protein would therefore be truncated, lacking both the dsRBDs and the catalytic deaminase motif. It is also possible that the premature stop codon triggers nonsense-mediated decay (NMD), rapidly degrading the RNA so that no protein is produced. Thus, in this case, editing could have the function of downregulating ADAR2 protein expression by alternative splicing.

In *Drosophila*, some transcripts of the single dADAR protein also undergo selfediting. In the adult-specific 3/4 splice version, a codon change from serine (S) to glycine (G) is introduced close to the zinc-chelating motif in the deaminase domain. The S/G exchange leads to the production of an enzyme with reduced catalytic activity (Keegan et al. 2005). Interestingly, self-editing of the 3/4 transcript increases throughout development, being lowest in embryos and highest in adult flies. Down-regulation of editing activity seems essential, since ubiquitous overexpression of a self-editing-deficient, hyperactive ADAR in *Drosophila* causes lethality (Keegan et al. 2005). Thus, auto-editing can also regulate the catalytic activity of ADARs throughout development.

5.5   GABA Receptor Editing

The most recent addition to edited receptors in the nervous system is the $GABA_A$ receptor subunit $\alpha3$ (Ohlson et al. 2007). The $GABA_A$ receptor is responsible for most inhibitory postsynaptic signals in the mammalian brain. These postsynaptic receptors are composed of five subunits with an extracellular ligand-binding domain and ion-channel domains that are integral to the membrane. The $GABA_A$ receptor subunits can be subdivided into seven families with multiple isoforms, but most receptors are believed to consist of $\alpha_{(1-6)}$, $\beta_{(1-3)}$, and $\gamma_{(1-3)}$ subunits in the ratio 2:2:1. The mammalian Gabra-3 transcript coding for the $\alpha3$ subunit is edited at a single site in transmembrane domain 3. A to I editing at this site causes an amino acid change from isoleucine to methionine, but the function of this editing event has not yet been elucidated. However, this site is developmentally regulated with a low level of editing during embryogenesis, increasing to almost 100% editing in the adult mouse brain. This, and the fact that another receptor involved in neurotransmission is edited make it likely that this site of editing is important for the function of the receptor.

5.6   Hepatitis Delta Virus (HDV) RNA Editing

There is only one example known where a virus utilizes site-selective A to I editing to increase the variety of proteins produced from the virus genome. The human hepatitis delta virus (HDV) has been shown to be edited at a single adenosine, converting an amber stop codon into a tryptophan (W) coding triplet (Luo et al. 1990; Casey and Gerin 1995). This amber/W site enables the virus to express a short and a long form of the viral protein delta antigen (HDAg), the only protein encoded by the virus. Both isoforms are required for the viral life cycle. The short form (HDAg-S) is used during viral replication (Kuo et al. 1989), while the long form (HDAg-L) is involved in packaging of new virus particles (Chang et al. 1991). HDV is a virus with a negative-strand RNA genome of about 1,700 nucleotides that replicates through a positive-strand intermediate called the antigenome. Editing occurs on the antigenome, and it has been shown that it is primarily ADAR1 that is responsible for this editing event (Jayan and Casey 2002).

5.7   Other Candidates for A to I Editing

Another substrate for editing has been found relatively recently in the pre-mRNA coding for the endothelin B receptor (ETB). In this case, editing by adenosine deamination was found in healthy individuals, although a much higher frequency of editing was found in patients with Hirschsprung disease (Tanoue et al. 2002). A to I editing was detected at position 950 in exon 4 of the ETB RNA. This editing event results in an amino acid substitution from glutamine (Q) to arginine (R) after translation. Editing at the Q/R site of ETB was observed only in transcripts that were subjected to alternative splicing by exon-skipping of exon 5. It is possible, but not yet proven, that the processes of alternative splicing and editing influence each other. Exon-skipping of exon 5 leads to a frame shift, which results in premature stop codons. It has been suggested that the alternative protein or mRNA is quickly degraded, since the product cannot be detected.

The human hematopoietic cell phosphatase (PTPN6) gene codes for a cytoplasmic protein expressed primarily during the development of hematopoietic cells. PTPN6 down-regulates a broad range of growth-promoting receptors by dephosphorylation. PTPN6 pre-mRNA is edited at several sites, with the main A to I conversion at A (7866), the putative branch point in intron 3 (Beghini et al. 2000). The editing event leads to aberrant splicing, resulting in retention of intron 3. Retention of the 251 nucleotides is predicted to change the open reading frame of the translated protein, creating a premature stop codon within the retained intron. Edited sites in the PTPN6 transcript were found in patients with acute myeloid leukemia. PTPN6 is believed to play a tumor-suppressing role, while alterations in PTPN6 expression have oncogenic potential.

A bioinformatics screen has recently identified four additional editing targets in mammals that are located in coding regions. Two of these were also detected in chicken. One of the substrates encodes filamin alpha (FLNA), a cytoskeletal protein that crosslinks actin filaments. Human FLNA consists of tandemly arranged IgG-like domains. Editing can be found in repeat number 22, but the biological consequence of the editing event remains to be determined. Editing of FLNA mRNA was detected in human, mouse, and chicken cDNAs. Edited transcripts are most abundant in brain but can also be found in other tissues (Levanon et al. 2005). The mRNA encoding CyFIP2, cytoplasmic FMR-1 interacting protein 2, is also edited in humans, mice, and chicken. Editing leads to a lysine (K) to glutamic acid (E) exchange, and can be detected only in the brain. BLCAP is a protein of unknown function that is highly conserved amongst all metazoans. Editing could be detected only in mammalian Blcap RNAs, leading to a tyrosine (Y) to cystein (C) exchange in the resulting protein (Clutterbuck et al. 2005; Levanon et al. 2005). IGFBP7 is an IGF-binding protein of unknown function. Editing of the mRNA encoding this protein was detected at two sites in mammals, leading to arginine (R) to glycine (G), and lysine (K) to arginine (R) substitutions. Analyses of RNAs derived from mice deficient in ADAR2 or ADAR1 suggest that CYFIP is edited predominantly by ADAR2, while FLNA can be edited by either enzyme. Finally, BLCAP is edited primarily by ADAR1 (E. Riedmann and M. Jantsch, unpublished data). However, the biological significance of these editing events remains enigmatic at this point.

Experimental and bioinformatics approaches have recently identified miRNAs as substrates for ADAR-mediated editing (Luciano et al. 2004; Blow et al. 2006). Editing of these RNAs might interfere with passenger strand selection and target preference. Interestingly, substrate sites in miRNAs seem to follow a certain consensus sequence (Blow et al. 2006). It could also be shown that miRNA editing interferes with Drosha and Dicer cleavage and thus maturation of the miRNA (Yang et al. 2006; Kawahara et al. 2007b).

5.8   Hyperediting and 3′ UTR Editing

Hyperediting, or non-selective editing, is found predominantly in viral RNAs. The editing patterns observed in hypermutated viral RNAs are consistent with ADARs being the enzymes responsible for the event, but direct experimental evidence is still lacking in most cases. Hyperediting of viral genomes is generally seen as a cellular antiviral defense mechanism. For instance, hyperediting can lead to nuclear retention of viral RNA (Kumar and Carmichael 1997). However, hyperediting of measles virus RNA can most likely also lead to persistent viral infection (Cattaneo 1994). Consistent with the proposed antiviral defense function of ADARs is the discovery of an interferon-inducible ADAR1 version in mammals. The interferon-inducible isoform of ADAR1 contains both nuclear import and export signals, and continuously shuttles between the nucleus and cytoplasm, thus being potentially able to edit viral RNAs in both cellular compartments (Poulsen et al. 2001; Strehblow et al. 2002; Desterro et al. 2003).

With the advent of systematic screens for inosines in RNAs, abundant editing was detected in non-coding regions of RNAs such as introns and 3′ untranslated regions. Biochemical detection of inosines originally identified editing in extensively structured 3′ UTRs in nematodes, but also in the mammalian brain (Morse and Bass 1999; Morse et al. 2002). More recently, several bioinformatics approaches have led to the identification of abundant editing in Alu elements in the human transcriptome (Athanasiadis et al. 2004; Blow et al. 2004; Kim et al. 2004; Levanon et al. 2004). Depending on the algorithm, and the stringency of the screening procedure used, between 13,000 and 30,000 edited sites were predicted in 1,600 to 2,600 genes. The vast majority of these sites were found in repetitive Alu elements, which are the most abundant short interspersed elements (SINEs) in the human genome. Editing typically occurs in pre-mRNAs, or RNAs that contain multiple Alu elements inserted in opposite orientations in relatively close vicinity (Athanasiadis et al. 2004). Base-pairing of these elements provides the double-stranded structure required for editing. Within each element, editing is moderate – in human brain, approximately one editing event can be found every 1,000 base-pairs of non-coding RNA (Blow et al. 2004).

The function of these editing events is as yet not clear. However, as pointed out below, inosines in RNA can affect the stability and localization of RNAs. Moreover, the secondary structures formed between repetitive elements could also trigger short interfering RNA (siRNA)-mediated silencing effects, which can also be adversely affected by ADARs. In any case, editing of repetitive elements is poorly conserved, and shows a high degree of species-specific variation. Alu elements, for instance, are primate-specific, while mice have a different predominant class of SINEs that are less conserved amongst themselves. Therefore, editing of transcribed SINEs is at least 30-fold lower in mice, and evidently occurs at different positions than in humans (Kim et al. 2004; Eisenberg et al. 2005). Taken together, this indicates that repetitive element editing might be a by-product of ADAR activity, rather than a highly specific regulatory mechanism. Alternatively, repetitive element editing might provide a defense mechan ism to safeguard the genome against massive retro-transposition events (Nishikura 2004).

6 Biological Roles of Editing and Animal Model Systems to Study Editing

Similarly to other alternative RNA-processing events, A to I editing has the potential to increase the variability of the transcriptome (Fig. 5). Since inosine is seen as guanosine by the cellular machinery, editing causes a functional A to G change. Within coding sequences, this can lead to codon substitutions with an effect on the translated protein. In turn, this may result in more than one isoform of the protein being expressed from one gene. A to I editing also has the potential to affect other RNA-processing events, such as pre-mRNA splicing, polyadenylation, and degradation. Several examples have been found where editing causes alternative splicing, either creating or disrupting sequences required for splice-site selection (Rueter et al. 1999; Beghini et al. 2000). One example of effects on polyadenylation has recently been suggested, where editing is involved in regulating the switch from early- to late-phase polyomavirus infection. The polyadenylation site of both the early and the late primary transcripts are edited. This leads to degradation of the early transcript, while the late transcript is stabilized (Gu et al. 2007).

Another potential role for A to I editing is to regulate RNA stability. By inducing alternative splicing, editing has the potential to stimulate mRNA degradation by the NMD pathway. Thus editing can regulate the expression of genes by controlling transcript stability. Similarly, inosines in RNA can lead to both specific degradation and nuclear retention of RNAs, thereby also affecting gene expression.

Site-selective A to I editing is a mechanism used to fine-tune the transcriptome, and increase the variety of protein isoforms expressed. This modification can be

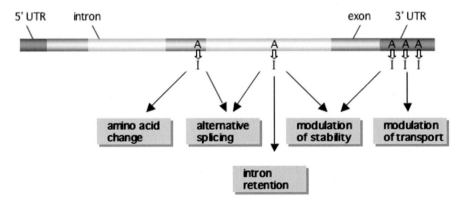

Fig. 5 Possible effects of A to I editing in the cell nucleus. Schematic overview of the occurrence of editing sites in a pre-mRNA, and their potential consequences. Editing events in exons (*green*), introns (*grey*), and untranslated regions (UTR, *red*) are shown. See text for further details

tissue-specific, and therefore be used in areas that require increased proteome diversity. One area that is known to require extensive protein diversity is the brain. It is well known that RNA processing such as alternative splicing is extensive and highly regulated in the nervous system (for a review, see Grabowski and Black 2001). It is therefore not surprising that pre-mRNA editing by adenosine deamination is found mainly in the central nervous system.

6.1   The Biological Role of ADAR1 and ADAR2

ADAR1 and ADAR2 have overlapping, but distinct substrate recognition abilities. For example, glutamate receptor subunit B pre-mRNA is edited mainly by ADAR2 at the Q/R site, while both ADAR1 and -2 can edit the R/G site. Another substrate that is edited by both enzymes is the serotonin receptor 5-HT_{2C} pre-mRNA. Here, the five edited sites are edited to different extents by the ADAR proteins. ADAR1 was shown to selectively edit two of the five edited sites in the serotonin receptor (Hartner et al. 2004).

Both ADAR1 and ADAR2 have been found to be essential in mammals. Homozygos null alleles of ADAR1 cause embryonic lethality in mice between days 11.5 and 12.5 (Hartner et al. 2004; Wang et al. 2004). The phenotypes of the null alleles include fast liver degradation, and severe defects in hematopoiesis. It is not known whether the phenotype results from altered editing at specific sites, or from lack of editing at non-specific sites in a more general pathway. ADAR2-deficient mice are viable. However, these mice suffer from epileptic seizures, and die between postpartum days 0 and 20 (Higuchi et al. 2000). The Q/R site editing in pre-mRNA transcripts of GluR-B from brain RNA was tenfold lower in $\text{ADAR2}^{-/-}$ animals than in wild-type mice, decreasing from 98 to 10%. Interestingly, $\text{ADAR2}^{-/-}$ animals are completely rescued, and appear to be normal, when the wild-type GluR-B allele is replaced by a GluR-B allele coding for the edited R codon at the Q/R site. Lack of Q/R site editing is therefore suggested to be the major cause of the $\text{ADAR2}^{-/-}$ phenotype. This is also in line with results showing that heterozygous mice synthesizing GluR-B that is not edited at the Q/R site show a similar phenotype to that of the $\text{ADAR2}^{-/-}$ animals (Brusa et al. 1995). Examples of other sites with a dramatic decrease in editing in $\text{ADAR2}^{-/-}$ animals are the GluR-B R/G site, the 5-HT_{2C} site C, the edited sites in ADAR2 pre-mRNA, and Gabra-3. Although not essential for survival, these edited sites presumably have a role in fine-tuning the transcriptome.

6.2   The Biological Role of Editing in Neurotransmission

Most of the genes targeted for adenosine deamination by ADARs are involved in neurotransmission. This is true for different targets in a variety of organisms, including *Drosophila*, squid, and humans. Above, we have discussed GluR-B and the essential editing at the Q/R site for the regulation of Ca^{2+} influx. A more recent function for the edited Q/R site of GluR-B concerns the maturation and assembly of GluR-B-containing AMPA receptors. It has been shown that the Q/R edited state of GluR-B has a negative effect on the tetramerization of the receptor during assembly. This results in a limited incorporation of Q/R-edited GluR-B into the AMPA receptor, and retention of the GluR-B subunit in the endoplasmatic reticulum (Greger et al. 2002, 2003). In this way, the Q/R site influences the properties of the receptor at two levels during ion conduction – by controlling the Ca^{2+} influx, and by restricting the number of AMPA receptors assembled with a GluR-B subunit.

Members of the serotonin receptor gene family 5-HT_2 are thought to play important roles in physiological and behavioral processes such as feeding behavior. mRNA of receptor subtype 5-HT_{2C} that is edited alters the coding potential of the second intracellular loop, reducing the efficiency of the interaction between the receptor and the G protein (Burns et al. 1997). Editing at the C and E site of the 5-HT_{2C} pre-mRNA that gives rise to an amino acid change at position 158 from asparagine to glycine has been shown to be regulated by serotonin. Serotonin-depleted mice exhibit decreased editing at these two sites. As a result, 5-HT_{2C} mRNA encoding receptors with a higher affinity for serotonin are expressed (Gurevich et al. 2002). In humans, isoforms carrying glycine at position 158 showed the strongest decrease in G-protein coupling activities (Wang et al. 2000). This result indicates that editing 5-HT_{2C} pre-mRNA is a way of regulating receptor response to the level of synaptic input.

6.3 Editing in Invertebrates

Deletion of the single *Drosophila* ADAR gene leads to severe behavioral defects, locomotion defects, and deterioration of the nervous tissue (Palladino et al. 2000a, b). Editing of the cacophony (*cac*) and paralytic (*para*) transcripts probably represents the most extensively studied examples of editing in the fly (Hanrahan et al. 2000; Palladino et al. 2000a). The two genes encode subunits of calcium and sodium channels, respectively, and editing of these channel subunits occurs at 11 and 10 sites, respectively. Flies mutant in these receptors reflect some, but not all of the defects described in dADAR-deficient flies, suggesting that other targets are responsible for the severe phenotypic effects observed in dADAR-deficient flies. In fact, using a bioinformatics approach, several other *Drosophila* editing targets have recently been identified (Hoopengardner et al. 2003). Biochemical screens have also led to the isolation of novel putative editing targets in *Drosophila* (Xia et al. 2005).

In *C. elegans*, two ADAR proteins have been identified. Deletion of the *adr-1* gene strongly reduces editing activity, while truncation of the *adr-2* gene seems to completely abolish editing activity. Interestingly, *adr-1*, *adr-2* double-mutants show severe chemotaxis defects, consistent with the neuronal expression pattern of these genes. While single mutants of *adr-1* show only relatively mild phenotypic effects, *adr-2* mutant animals are strongly affected in chemotaxis. Most interestingly, a link between RNAi and editing was established in *C. elegans*, and has meanwhile also been demonstrated in vertebrates (see below).

6.4   Does Editing Affect RNA Degradation?

In the pre-mRNA of ADAR2 and PTPN6, editing causes alternative splicing that leads to a frame shift. The change of open reading frame creates a premature stop codon that might lead to the expression of a truncated protein. As mentioned above, another possibility is that the RNA is subjected to NMD. It is believed that the mammalian cell routinely utilizes alternative splicing to trigger NMD in order to regulate protein levels (reviewed in Lejeune and Maquat 2005). It is therefore possible that editing-induced alternative splicing is also used to regulate gene expression at the RNA level. Furthermore, in a recent report, the NMD pathway was down-regulated by siRNA-mediated depletion of hUpf1 in HeLa cells (Mendell et al. 2004). A twofold induction of ADAR2 mRNA was observed by microarray analysis, comparing depleted with undepleted cells. This result indicates that the ADAR2 transcript is stabilized in NMD-deficient cells. In line with this observation, less than 5% of the ADAR2 proteins correspond to the truncated protein, although 80% of ADAR2 transcripts were subjected to editing and subsequent alternative splicing in rat brain (Rueter et al. 1999). In summary, this might indicate that the editing-induced alternatively spliced ADAR2 pre-mRNA is subjected to NMD.

6.5   Editing and Disease

Over the past few years, several studies have investigated correlations between changes in editing patterns of certain, predominantly neuronal ADAR substrates and specific disease conditions. Probably most attention has been given to editing of the serotonin 5-HT_{2C} receptor encoding RNA, since this receptor is a major target for antidepressants. As mentioned above, editing of this receptor strongly influences its G-protein coupling efficiency. In fact, interferon, which stimulates the expression of the long cytoplasmic p150 version of ADAR1, also leads to altered editing of the serotonin 5-HT_{2C} receptor, which could be a possible explanation for the observed induction of depression in patients on interferon treatment (Yang et al. 2004). In addition, a rat model system for depression recently demonstrated alterations of 5-HT_{2C} editing patterns in response to antidepressants (Iwamoto et al. 2005). Thus, the editing status of this receptor might well have psychopathological consequences.

Alterations of editing patterns in human malignant gliomas have also been described. Here, reduced editing of the Q/R site in the GluR-B subunit encoding RNA was observed, which is consistent with a reduction in ADAR2 activity (Maas et al. 2001).

Similarly, a reduction of editing at the Q/R site of GluR-B was observed in motor neurons of patients suffering from amyotrophic lateral sclerosis (Kawahara et al. 2004). Nonetheless, it is not yet clear whether the observed correlations between altered editing patterns and certain human diseases have a causative relation to disease development, or are merely an effect of the latter.

7   Function of Inosines in Non-Coding Sequences

While site-specific editing in coding substrates clearly increases the proteomic diversity of the organism by allowing the production of functionally distinct proteins from a given gene, the biological role of multiple editing events in non-coding regions is less clear. However, several observations indicate that editing can influence the stability and translatability of RNAs.

In *Xenopus*, specific binding of a protein to poly-inosine-containing RNAs was observed in the nucleus. The corresponding mammalian protein, termed p54nrb, can bind inosine-containing RNAs in a complex with the nuclear matrix protein matrin3 and splice factor PSF, leading to nuclear retention of hyperedited RNAs (Zhang and Carmichael 2001). In contrast, RNAs specifically edited at a single or a few sites are bound with low affinity by this complex, and can therefore be exported from the nucleus. Nuclear retention of hyperedited RNA was thus seen as a mechanism to safeguard the quality of an RNA, or to regulate RNA levels via hyperediting, possibly by antisense transcription (Zhang and Carmichael 2001).

Another function for the nuclear retention of mRNA has been proposed by Prasanth et al. (2005). The <u>C</u>AT2 <u>t</u>ranscribed <u>n</u>uclear-RNA (CTN-RNA) is transcribed from the <u>m</u>ouse <u>c</u>ationic <u>a</u>mino acid <u>t</u>ransporter <u>2</u> (mCAT2) gene, using an alternative promoter and poly(A) site. The CTN-RNA has an extended 3′ UTR, compared with the mCAT2 mRNA. The CTN-RNA has been proposed to be A to I edited at several sites in the 3′ UTR, possibly because inverted repeats form double-stranded structures in this region (Prasanth et al. 2005). Due to interactions with p54nrb and PSF, the hyperedited RNA is retained in the nucleus. Under stress conditions, the CTN-RNA is post-transcriptionally cleaved, producing a shorter mRNA containing the mCAT coding region, but lacking the 3′ UTR where the edited sites are situated. The CTN-RNA has therefore been suggested to regulate the protein levels of its corresponding protein-coding partner.

The heterochromatin-associated protein vigilin has also been found to associate with inosine-containing RNA (Wang et al. 2005). Vigilin and its *Drosophila* homolog DDP1 also co-localize with heterochromatin protein HP1, and experiments indicate that vigilin is required for proper localization of HP1 and thus proper heterochromatin formation (Huertas et al. 2004). Interestingly, when bound to inosine-containing RNA, vigilin was also found associated with DNA PK, indicating a possible role of this complex in the DNA repair pathway. However, while it is clear that ADARs interfere with the RNAi machinery (see below), and also that the RNAi machinery is involved in heterochromatin formation, it still remains to be proven that inosine-containing RNA is required for either heterochromatin formation or DNA repair. Along these lines, it is also worth mentioning that *Xenopus* ADAR1 can be found highly enriched at a unique chromosomal region (Eckmann and Jantsch 1999). This site is transcriptionally silent, and localization of ADAR1 to this chromosomal domain does not require ongoing transcription. However, digestion with RNAses removes ADAR from this locus, suggesting that an RNA molecule tethers ADAR1 to this site. The function of this chromosomal site is not entirely clear, but it seems possible that chromosomal tethering of

ADAR1 in *Xenopus* might regulate the intranuclear concentration of ADAR1 (see below). In any case, this finding demonstrates that ADAR1 can associate with chromatin, possibly in an RNA-bound fashion.

RNAs containing inosines can also be specifically degraded. A nuclease that recognizes inosines in RNA can be found in several tissues. This cytoplasmic, 3′–5′ nuclease preferentially cleaves single-stranded RNAs, but can also degrade RNAs modified by ADAR2 at sites of alternating IU or UI base-pairs (Scadden and Smith 1997, 2001b). Affinity purification of this nuclease has identified Tudor staphylococcal nuclease (TSN) as one component of this complex. While TSN can bind to inosine-containing RNAs and stimulate their cleavage, it is not sufficient for cleavage, indicating that other components of the inosine-specific nuclease are still to be identified. Most interestingly, TSN is a component of the RISC (Caudy et al. 2003). dsRNAs can obviously follow one of two pathways: the RNA can either be deaminated by ADARs, or serve as triggers for the RNAi pathway. It has been shown that these two pathways compete with each other (see below). It therefore seems possible that TSN prevents deaminated inosine-containing RNAs from entering the RNAi pathway by specific binding, thus marking them for degradation (Scadden 2005).

It is becoming increasingly evident that RNA editing and RNAi are competing with each other. In nematodes, for instance, transgene-induced silencing is strongly enhanced in the absence of ADARs, suggesting that ADARs can unwind dsRNA and thereby interfere with the formation of siRNAs (Knight and Bass 2002). Consistently, the aforementioned chemotactic phenotypes exhibited by ADAR mutant worms can be partially rescued by mutations in the RNAi pathway, supporting the idea that ADARs prevent or antagonize the feeding of structured or dsRNAs into the RNAi pathway (Tonkin and Bass 2003). In fact, in vitro experiments performed with *Drosophila* extracts indicate that extensive editing by ADARs inhibits siRNA formation. However, moderately edited dsRNAs can be converted into siRNAs by Dicer, giving rise to siRNA-containing inosines (Scadden and Smith 2001a). Since inosine-containing siRNAs will be impaired in their base-pairing potential with target RNAs, it is likely that edited siRNAs will have a strongly reduced efficiency in triggering RNAi. Alternatively, such modified RNAs might be degraded by the aforementioned TSN complex.

Furthermore, strong binding of the long, cytoplasmic p150 form of ADAR1 to siRNAs has been reported. Mouse embryonic fibroblast cells lacking ADAR1 also showed an increased response to RNAi, while overexpression of ADAR1 led to a reduced RNAi response (Yang et al. 2005). Thus, ADARs might not only prevent the feeding of dsRNAs into the production of siRNAs, but may also interfere with their activity by generating inosine-containing siRNAs, possibly prone to degradation. The binding of ADAR to siRNAs in the cytoplasm might also interfere with the formation of the RISC complex.

Besides siRNAs, microRNAs can also become edited (Luciano et al. 2004; Blow et al. 2006). Edited miRNAs can be prevented from proper processing and are degraded, possibly by TSN (Yang et al. 2006). Moreover, editing can redirect the microRNA to target a different set of genes than for the unedited microRNA (Kawahara et al. 2007a).

The human microRNA miR376a-5p is edited at one site in the seed sequence, critical for the microRNA to hybridize to the target molecule. Approximately 80 target genes were predicted for both the edited and the unedited form of miRNA-376a-5p. However, only two of the target genes were common between the two forms. This result indicates that editing also has an influence on gene regulation by RNA interference.

8 Structures of Specific Substrates

The inverted repeats, creating the double-stranded editing substrate, can be separated by a fairly large unstructured region. There are examples where the paired regions are separated by more than 1,000 nucleotides, but the loop separating the inverted repeats can also be as short as five nucleotides (Fig. 4), as in the case of the stem-loop structure surrounding the GluR-B R/G site. Due to its conveniently small size and efficient in vitro editing, this substrate of about 70 nucleotides has been used in several structure analyses.

Intron sequences are often used to generate the double-stranded structure required for editing within an exon (Fig. 4). It is not surprising that the editing complementary sequence is situated within an intron, as intronic sequences do not contribute to the translated protein, and are therefore more tolerant to base changes that can create an inverted repeat. However, there are exceptions. The intronless gene coding for the human $K_v1.1$ potassium channel is edited at one adenosine, and the editing complementary sequence is also situated within the encoded sequence (Hoopengardner et al. 2003).

Although most sites for editing are situated within the helix region of a stem loop, more complex structures have been found. One example is the transcript of the synaptotagmin I (*sytI*) gene in *Drosophila*, which is edited at four sites (A to D) in one of its exons (Reenan 2005). The double-stranded regions required to achieve editing are created through a pseudoknot structure. Reenan (2005) also showed that transcripts of the *sytI* gene were edited in other species. Fruitfly, mosquitoes, and butterflies have both shared and species-specific sites of editing, all within the same corresponding exon. Several of these species also produce pseudoknot structures of the edited transcript. Species specificity is created by small changes in intron sequences distant from the edited site, rather than by specificity acquired through the editing enzyme. The genesis of RNA-editing sites is proposed to have occurred by stepwise addition of structural domains, or by mutating a small number of bases in ancestral structures.

8.1 Site-Selective Editing

It is hard to predict what a good, selectively edited substrate for A to I editing should look like, as no specific consensus sequence has been defined for ADAR editing. This is, however, not so surprising, considering that dsRNA-binding enzymes such as ADARs often bind in a sequence-independent manner. Nonetheless, ADAR enzymes are influenced by the sequence surrounding the editing site. In vitro studies using artificial substrates have shown that ADAR1 preferentially targets an adenosine with a 5′ nearest neighbor of U or A; a C at this position is more seldom, and a G very rare (Polson and Bass 1994; Lehmann and Bass 2000). ADAR2 has a similar preference for the base 5′ of the edited site; U and A are preferable to C and G (Lehmann and Bass 2000). ADAR2 also has a preferred base 3′ of the edited site (U = G > C = A; reviewed in Bass 2002). Although these studies have been done on complete duplexes in vitro, they are in agreement with most sites that are edited in vivo. In an attempt to define structure and sequence determinants for ADAR2 substrates, Dawson et al. (2004) compared the ability of ADAR2 to edit four different site-selectively edited substrates at a number of different sites. This study, and previous studies, came to the conclusion that both sequence and structure determine a good editing site. The structure of most site-selectively edited substrates consists of a stem loop that is up to 80 base-pairs long, interrupted by internal loops and mismatches. The deaminated adenosine does not necessarily have to be base-paired; commonly, there is a C opposite the A. ADAR1 has even been shown to increase its editing efficiency at the GluR-B Q/R site upon a change from an AU base-pair to a mismatched AC (Maas et al. 1996). In the mammalian brain, GluR-B is edited mainly by ADAR2 at the Q/R site. In fact, an AC mismatch is preferred to AU base-pairs for efficient editing in most cases (Wong et al. 2001). However, other mismatched nucleotides, such as A or G, have never been recorded opposite an edited site. Mutational analysis has shown that a G opposite the R/G editing site in GluR-B affects both selectivity and efficiency of editing (Källman et al. 2003). It is possible that the more bulky purines block deamination. ADAR enzymes have been proposed to use a base-flipping mechanism for catalysis, similar to that described for nucleoside deaminases and DNA methyltransferases (Allan and Reich 1996; Hough and Bass 1997). Increased conformational flexibility of the bases opposite the R/G site during ADAR2 binding has been observed, indicating that a base-flipping mechanism is used (Yi-Brunozzi et al. 2001).

Another potential function of mismatched regions in site-selectively edited substrates is that they direct the enzyme to bind to the specific site of editing. Footprinting analysis has shown that GluR-B RNA has a preference for ADAR2 to be positioned at the R/G editing site (Öhman et al. 2000). The binding site in the R/G stem loop is a region of 11–16 bases surrounding the R/G site. A mutant substrate, lacking two internal loops in the R/G stem, has a similar affinity for ADAR2 but the binding within the stem is non-selective. Furthermore, a prominent negative effect on R/G site selectivity, allowing other adenosines to be edited, was shown in a mutant transcript where the 31-base-pair-long R/G stem was made into a perfect helix. This suggests that site-selective editing occurs, at least partly, at the level of binding, and that internal loops in the stem of the hairpin are involved in recognition.

In general, site-selectively edited substrates consist of double-stranded regions interrupted by internal loops and mismatching bulges, while hyperedited substrates modified at many sites have longer stretches of almost completely double-stranded structure. However, the borders between a typical site-selectively edited substrate and a hyperedited one cannot be drawn easily. Helices of substrates that are considered to be hyperedited are sometimes also interrupted by bulges and loops. In an attempt to investigate whether ADAR2 can distinguish between hyperediting of a long, completely base-paired duplex and a selectively edited site such as the GluR-B R/G site, a molecule containing both of these structures was constructed. Using scanning force microscopy on this substrate in the presence of ADAR2, it was possible to distinguish whether ADAR2 preferred to bind to a specific target that is site-selectively edited over a random complete duplex of several hundred base-pairs (Klaue et al. 2003). Indeed, ADAR2 preferentially bound at the site-selectively edited R/G site in over 90% of the molecules.

9 Regulation of Editing

ADAR-mediated editing is tightly controlled at several levels. Most strikingly, self-editing of ADARs, and the use of alternative splice sites are central mechanisms regulating editing activity, common to several organisms ranging from *Drosophila* to man.

As mentioned above, the rat ADAR2 enzyme can edit its own pre-mRNA in intron 4 (Rueter et al. 1999). The AI di-nucleotide created after editing mimics the AG sequence required for an acceptor splice site, and serves as an alternative 3′ splice site. The 47 nucleotides included in the alternative exon change the open reading frame, producing premature stop codons within alternative exon 5. Thus, it has been suggested that A to I editing of ADAR2 pre-mRNA functions as an auto-regulatory feedback loop to control ADAR2 levels, and thereby editing activity. Up to 80% of all ADAR2 pre-mRNAs extracted from whole rat brain are edited, and subsequently alternatively spliced at this site (Rueter et al. 1999).

9.1 Possible Regulation of ADAR1 by RNA Interference

Although ADAR1 is expressed in most tissues, editing levels vary greatly amongst different tissues. This could be partially due to a regulation of the ADAR1 transcript. A microRNA (miR-1) targets the untranslated region (UTR) of human ADAR1 (Lim et al. 2005). Tissue-specific expression of miR-1 was observed, and revealed increased expression in heart and skeletal muscle. Furthermore, the miR-1 repressed a reporter gene bearing the 3′ UTR segment of ADAR1. One possibility could therefore be that the expression of ADAR1 is suppressed by the RNAi machinery in certain tissues.

9.2 Transcriptional Regulation

Transcriptional control of ADARs has been best studied in mammals that have three ADAR proteins, ADAR1, −2, and −3. The basic expression profile of these three proteins varies considerably. While ADAR2 is most prominently expressed in the brain, ADAR1 shows a more general expression pattern. ADAR3 is restricted to the brain (Kim et al. 1994; Melcher et al. 1996a, b).

ADAR1 can be expressed from two different promoters, and shows several alternative splicing patterns (Liu et al. 1997; George and Samuel 1999). One of the promoters is constitutive, while the other is interferon-induced. The constitutive promoter gives rise to a shorter ADAR1 protein that starts at Met 296. This shorter protein lacks the ZBDs and the NES located in the amino terminus of the protein, and therefore localizes to the nucleus. The interferon-induced version of the protein is longer, and due to the inclusion of the NES, this version of the protein is predominantly cytoplasmic, and might primarily serve to edit viral RNAs in the cytoplasm (Patterson and Samuel 1995; Eckmann et al. 2001; Poulsen et al. 2001).

9.3 Intracellular Localization

Differential intracellular localization might represent another mechanism of regulating editing activity in certain cellular compartments. As pointed out above, the interferon-induced long version of ADAR1 is predominantly cytoplasmic, but continuously shuttles between the nucleus and the cytoplasm (Desterro et al. 2003). Constitutively expressed ADAR1 can also sometimes be found in the cytoplasm (Strehblow et al. 2002; Yang et al. 2003). Moreover, both ADAR1 and ADAR2 can accumulate in nucleoli (Desterro et al. 2003; Sansam et al. 2003). Injection of substrate RNAs into cells releases the enzymes from nucleoli, suggesting that the enzymes are stockpiled in this nuclear organelle, possibly to sequester the enzyme, thereby preventing inadvertent editing of endogenous RNAs (Sansam et al. 2003).

Xenopus ADAR1 can also be found enriched in a particular, transcriptionally silent chromosomal region. Injection of substrate RNAs into oocyte nuclei can efficiently strip the enzymes off this site, suggesting that ADAR1 is stored at this site, possibly to regulate the intranuclear enzyme activity (Sallacz and Jantsch 2005).

Specific editing by ADAR2 has also been reported to occur in the nucleolus. When RNAs containing the GluR-B Q/R site, or the serotonin 5-HT$_{2C}$ receptor editing sites were targeted to the nucleolus by expression from a Pol I promoter, editing of the Q/R site in GluR-B, and of sites C and D in 5-HT$_{2C}$ were specifically edited upon co-transfection of ADAR2, but not ADAR1 (Vitali et al. 2005).

Yet another mechanism regulating ADAR-mediated RNA editing might exist in nucleoli. The C/D box snoRNA MBII-52 is complementary to the edited region of

5-HT$_{2C}$ RNA, and predicted to target editing site C for ribose methylation, which in turn could down-regulate editing by ADARs. In fact, overexpression of MBII-52 snoRNA can adversely affect editing of site C in a serotonin receptor 5-HT$_{2C}$ construct that is artificially targeted to nucleoli (Vitali et al. 2005). Thus, targeted methylation of editing sites by C/D box snoRNAs might be a mechanism by which editing efficiencies of editing regions located in close proximity might be fine-tuned.

9.4   Coordination of Editing and Pre-mRNA Splicing by the Transcriptional Machinery

Most A to I editing sites in mammalian pre-mRNAs are situated in the vicinity of exon–intron or intron–exon borders. Intron sequences complementary to the edited sites create the dsRNA structure required for modification. Thus, only the pre-mRNA can act as a substrate in vivo, and editing has to occur during a relatively short time window before the splicing event. The C-terminal domain (CTD) of the largest subunit of RNA polymerase II (Pol II) has been proposed to be the main coordinator of RNA-processing events (reviewed in Zorio and Bentley 2004; Kornblihtt et al. 2004). The CTD promotes RNA processing by capping, splicing, and cleavage/polyadenylation in mammalian cells. The phosphorylated CTD, but not the unphosphorylated one, can bind to and co-localize with splicing factors (Mortillaro et al. 1996; Neugebauer and Roth 1997). Analysis of in vitro editing and splicing of the GluR-B transcript in the vicinity of the R/G site has revealed that editing and splicing can interfere with each other when their sites of recognition are in proximity to each other (Bratt and Öhman 2003). A model for synchronized editing and splicing in vivo, in which the CTD coordinates editing of the glutamate receptor pre-mRNA with splicing, has been proposed. The coordination between ADAR2 editing, splicing, and the role of the CTD as a synchronizer of these two RNA-processing events have also been investigated. Studies of ADAR2 and GluR-B pre-mRNA show that the CTD is required for coordination when sites of editing and splicing are in close proximity to each other, like at the R/G site (Laurencikiene et al. 2006; Ryman et al. 2007). Furthermore, in the GluR-B Q/R RNA substrate, where the editing site is more distant to the splicing donor site, the CTD has a less prominent effect on editing. However, the coordination between editing and splicing seems to be disrupted without the CTD, causing an increase in splicing efficiency (Ryman et al. 2007). During the editing-induced alternative splicing of the ADAR2 transcript, it is crucial that splicing is restrained until editing takes place, and the new 3′ splice site is formed (Laurencikiene et al. 2006). Surprisingly, the CTD is not required for efficient splicing of this pre-mRNA; both normal and alternative splicing occur independently of the CTD. However, also in this substrate the CTD is required for efficient co-transcriptional editing of the pre-mRNA. These results indicate that RNA editing can depend on the CTD. Editing, in turn, can influence the efficiency of splicing. Constructs containing the R/G site and the adjacent intron 13 of GluR-B pre-mRNA are spliced less efficiently in the presence of ADAR2 than in its absence. Interestingly, the edited base at the R/G site is sufficient to lower splicing efficiency of the adjacent intron. In this case, the editing site is two bases upstream of the 5′ splice site, and it appears that editing lowers the efficiency at which this site is recognized by the splicing machinery (Schoft et al. 2007). Reducing the speed of splicing also has biological implication. Alternative splicing occurs downstream of the edited R/G site located at the very end of exon 13 in GluR-B RNA. Unedited pre-mRNAs show an increase in erroneous splicing patterns at this alternative splice site, suggesting that the reduced splicing efficiency observed after R/G site editing may be important for proper splice-site selection in this alternatively spliced region (Schoft et al. 2007).

A similar case has recently been reported in *Drosophila*, where a strong correlation between alternative splice-site selection and editing status was found in a gene where the editing site was located 5′ to the alternatively spliced region. In contrast, no correlation between alternative splicing and editing was observed when the editing site was located downstream of the alternatively spliced region (Agrawal and Stormo 2005).

Another interaction of splicing and editing can be observed at the Q/R site of GluR-B RNA. In contrast to the R/G site, the Q/R site is located 24 nucleotides upstream of the 5′ splice site. In vivo, this site is edited exclusively by ADAR2, with almost 100% efficiency. Most interestingly, in the absence of ADAR2 the adjacent intron fails to be removed, leading to nuclear retention of the unspliced RNA (Higuchi et al. 2000). Here, simultaneous editing of an intronic hotspot and the Q/R site is required for efficient splicing, indicating that editing can also stimulate the splicing reaction (Schoft et al. 2007).

9.5 Protein Modification

Protein modifications can modulate editing activity. Mammalian full-length ADAR1 (p150) is sumoylated in its amino terminus. Sumoylated ADAR1 is significantly less active than the unmodified protein, indicating that this type of protein modification can also regulate enzyme activity (Pinto Desterro et al. 2005).

9.6 Dimerization

Dimerization of ADARs might modulate editing efficiency. In mammals, ADAR1 and ADAR2 form homodimers, and even ADAR3 seems to exist in a dimeric form in brain extracts (Cho et al. 2003). Similarly, *Drosophila* dADAR also forms dimers via its first dsRBD (Gallo et al. 2003). It is noteworthy that dADAR requires RNA binding for successful dimer formation, while mammalian ADARs can homodimerize in the absence of RNA. Most interestingly, both mammalian and

Drosophila ADARs can from dimers with enzymatically inactive, or less active versions of the enzyme. Complexes formed between active and inactive ADARs show markedly reduced editing activity, compared with homodimers formed between active ADARs (Cho et al. 2003; Gallo et al. 2003). Since self-editing and alternative splicing events can generate ADARs of various activities both in mammals and in *Drosophila*, dimer formation between ADARs exhibiting different levels of enzymatic activity provides a fast regulatory mechanism to modulate editing activity in vivo.

10   Outlook

Until recently, RNA editing by ADARs was believed to be a rare and very specialized event that would affect only a few selected mRNAs. The findings that ADAR1-deficient mice die prematurely, and that several cellular and developmental functions are disrupted in the absence of ADARs have certainly changed this view, and led to a quest for the detection of novel editing targets. An example is the recent detection of abundant editing of transcribed repetitive elements in the mammalian transcriptome. With an increasing number of substrates that affect not only the translational code but also the other RNA-processing events, A to I editing has the potential to impart strong variation in the proteome, resembling the power of alternative splicing. Moreover, understanding the impact and potential regulatory role of ADAR editing, in particular with respect to crosstalk with the RNAi machinery, has given a new and exciting twist to the study of ADAR-mediated editing.

Acknowledgements The authors would like to thank Maria Siomos for critical review of the manuscript. Work in the labs of MFJ and MÖ is supported by the Austrian Science Foundation and the Swedish Research Council, respectively. We would like to thank all our colleagues for sharing unpublished data prior to publication.

References

Agrawal R, Stormo GD (2005) Editing efficiency of a *Drosophila* gene correlates with a distant splice site selection. RNA 11:563–566
Allan BW, Reich NO (1996) Targeted base stacking disruption by the EcoRI DNA methyltransferase. Biochemistry 35:14757–14762
Athanasiadis A, Rich A, Maas S (2004) Widespread A-to-I RNA editing of Alu-containing mRNAs in the human transcriptome. PLoS Biol 2:e391
Bass BL (2002) RNA editing by adenosine deaminases that act on RNA. Annu Rev Biochem 71:817–846
Bass BL, Weintraub H (1988) An unwinding activity that covalently modifies its double-stranded RNA substrate. Cell 55:1089–1098
Bass BL, Nishikura K, Keller W, Seeburg PH, Emeson RB, O'Connell MA, Samuel CE, Herbert A (1997) A standardized nomenclature for adenosine deaminases that act on RNA. RNA 3:947–949

Beghini A, Ripamonti CB, Peterlongo P, Roversi G, Cairoli R, Morra E, Larizza L (2000) RNA hyperediting and alternative splicing of hematopoietic cell phosphatase (PTPN6) gene in acute myeloid leukemia. Hum Mol Genet 9:2297–2304

Bhalla T, Rosenthal JJ, Holmgren M, Reenan R (2004) Control of human potassium channel inactivation by editing of a small mRNA hairpin. Nat Struct Mol Biol 11:950–956

Blow M, Futreal PA, Wooster R, Stratton MR (2004) A survey of RNA editing in human brain. Genome Res 14:2379–2387

Blow MJ, Grocock RJ, van Dongen S, Enright AJ, Dicks E, Futreal PA, Wooster R, Stratton MR (2006) RNA editing of human microRNAs. Genome Biol 7:R27

Bratt E, Öhman M (2003) Coordination of editing and splicing of glutamate receptor pre-mRNA. RNA 9:309–318

Brusa R, Zimmermann F, Koh DS, Feldmeyer D, Gass P, Seeburg PH, Sprengel R (1995) Early-onset epilepsy and postnatal lethality associated with an editing-deficient GluR-B allele in mice. Science 270:1677–1680

Burns CM, Chu H, Rueter SM, Hutchinson LK, Canton H, Sanders-Bush E, Emeson RB (1997) Regulation of serotonin-2C receptor G-protein coupling by RNA editing. Nature 387:303–308

Casey JL, Gerin JL (1995) Hepatitis D virus RNA editing: specific modification of adenosine in the antigenomic RNA. J Virol 69:7593–7600

Cattaneo R (1994) Biased (A–>I) hypermutation of animal RNA virus genomes. Curr Opin Genet Dev 4:895–900

Caudy AA, Ketting RF, Hammond SM, Denli AM, Bathoorn AM, Tops BB, Silva JM, Myers MM, Hannon GJ, Plasterk RH (2003) A micrococcal nuclease homologue in RNAi effector complexes. Nature 425:411–414

Chang FL, Chen PJ, Tu SJ, Wang CJ, Chen DS (1991) The large form of hepatitis delta antigen is crucial for assembly of hepatitis delta virus. Proc Natl Acad Sci USA 88:8490–8494

Chen CX, Cho DS, Wang Q, Lai F, Carter KC, Nishikura K (2000) A third member of the RNA-specific adenosine deaminase gene family, ADAR3, contains both single- and double-stranded RNA binding domains. RNA 6:755–767

Cho DS, Yang W, Lee JT, Shiekhattar R, Murray JM, Nishikura K (2003) Requirement of dimerization for RNA editing activity of adenosine deaminases acting on RNA. J Biol Chem 278:17093–17102

Clutterbuck DR, Leroy A, O'Connell MA, Semple CA (2005) A bioinformatic screen for novel A-I RNA editing sites reveals recoding editing in BC10. Bioinformatics 21:2590–2595

Dawson TR, Sansam CL, Emeson RB (2004) Structure and sequence determinants required for the RNA editing of ADAR2 substrates. J Biol Chem 279:4941–4951

Desterro JM, Keegan LP, Lafarga M, Berciano MT, O'Connell M, Carmo-Fonseca M (2003) Dynamic association of RNA-editing enzymes with the nucleolus. J Cell Sci 116:1805–1818

Doyle M, Jantsch MF (2003) Distinct in vivo roles for double-stranded RNA-binding domains of the *Xenopus* RNA-editing enzyme ADAR1 in chromosomal targeting. J Cell Biol 161:309–319

Eckmann CR, Jantsch MF (1999) The RNA-editing enzyme ADAR1 is localized to the nascent ribonucleoprotein matrix on *Xenopus* lampbrush chromosomes but specifically associates with an atypical loop. J Cell Biol 144:603–615

Eckmann CR, Neunteufl A, Pfaffstetter L, Jantsch MF (2001) The human but not the Xenopus RNA-editing enzyme ADAR1 has an atypical nuclear localization signal and displays the characteristics of a shuttling protein. Mol Biol Cell 12:1911–1924

Eisenberg E, Nemzer S, Kinar Y, Sorek R, Rechavi G, Levanon EY (2005) Is abundant A-to-I RNA editing primate-specific? Trends Genet 21:77–81

Fernandez HR, Kavi HH, Xie W, Birchler JA (2005) Heterochromatin: on the ADAR radar? Curr Biol 15:R132–134

Gallo A, Keegan LP, Ring GM, O'Connell MA (2003) An ADAR that edits transcripts encoding ion channel subunits functions as a dimer. EMBO J 22:3421–3430

Geiger JR, Melcher T, Koh DS, Sakmann B, Seeburg PH, Jonas P, Monyer H (1995) Relative abundance of subunit mRNAs determines gating and Ca2+ permeability of AMPA receptors in principal neurons and interneurons in rat CNS. Neuron 15:193–204

George CX, Samuel CE (1999) Human RNA-specific adenosine deaminase ADAR1 transcripts possess alternative exon 1 structures that initiate from different promoters, one constitutively active and the other interferon inducible. Proc Natl Acad Sci USA 96:4621–4626

Gerber A, Grosjean H, Melcher T, Keller W (1998) Tad1p, a yeast tRNA-specific adenosine deaminase, is related to the mammalian pre-mRNA editing enzymes ADAR1 and ADAR2. EMBO J 17:4780–4789

Grabowski PJ, Black DL (2001) Alternative RNA splicing in the nervous system. Prog Neurobiol 65:289–308

Greger IH, Khatri L, Ziff EB (2002) RNA editing at arg607 controls AMPA receptor exit from the endoplasmic reticulum. Neuron 34:759–772

Greger IH, Khatri L, Kong X, Ziff EB (2003) AMPA receptor tetramerization is mediated by Q/R editing. Neuron 40:763–774

Gu R, Zhang Z, Carmichael GG (2007) How a small DNA virus uses dsRNA but not RNAi to regulate its life cycle. Cold Spring Harbor Symp Quant Biol LXXI:1–7

Gurevich I, Englander MT, Adlersberg M, Siegal NB, Schmauss C (2002) Modulation of serotonin 2C receptor editing by sustained changes in serotonergic neurotransmission. J Neurosci 22:10529–10532

Hanrahan CJ, Palladino MJ, Ganetzky B, Reenan RA (2000) RNA editing of the *Drosophila* para Na(+) channel transcript. Evolutionary conservation and developmental regulation. Genetics 155:1149–1160

Hartner JC, Schmittwolf C, Kispert A, Muller AM, Higuchi M, Seeburg PH (2004) Liver disintegration in the mouse embryo caused by deficiency in the RNA-editing enzyme ADAR1. J Biol Chem 279:4894–4902

Herbert A, Rich A (2001) The role of binding domains for dsRNA and Z-DNA in the in vivo editing of minimal substrates by ADAR1. Proc Natl Acad Sci USA 98:12132–12137

Herbert A, Alfken J, Kim YG, Mian IS, Nishikura K, Rich A (1997) A Z-DNA binding domain present in the human editing enzyme, double-stranded RNA adenosine deaminase. Proc Natl Acad Sci USA 94:8421–8426

Herbert A, Schade M, Lowenhaupt K, Alfken J, Schwartz T, Shlyakhtenko LS, Lyubchenko YL, Rich A (1998) The Zalpha domain from human ADAR1 binds to the Z-DNA conformer of many different sequences. Nucleic Acids Res 26:3486–3493

Higuchi M, Maas S, Single FN, Hartner J, Rozov A, Burnashev N, Feldmeyer D, Sprengel R, Seeburg PH (2000) Point mutation in an AMPA receptor gene rescues lethality in mice deficient in the RNA-editing enzyme ADAR2. Nature 406:78–81

Hoopengardner B, Bhalla T, Staber C, Reenan R (2003) Nervous system targets of RNA editing identified by comparative genomics. Science 301:832–836

Hough RF, Bass BL (1997) Analysis of *Xenopus* dsRNA adenosine deaminase cDNAs reveals similarities to DNA methyltransferases. RNA 3:356–370

Huertas D, Cortes A, Casanova J, Azorin F (2004) Drosophila DDP1, a multi-KH-domain protein, contributes to centromeric silencing and chromosome segregation. Curr Biol 14:1611–1620

Iwamoto K, Nakatani N, Bundo M, Yoshikawa T, Kato T (2005) Altered RNA editing of serotonin 2C receptor in a rat model of depression. Neurosci Res 53:69–76

Jayan GC, Casey JL (2002) Inhibition of hepatitis delta virus RNA editing by short inhibitory RNA-mediated knockdown of ADAR1 but not ADAR2 expression. J Virol 76:12399–12404

Källman AM, Sahlin M, Öhman M (2003) ADAR2 A–>I editing: site selectivity and editing efficiency are separate events. Nucleic Acids Res 31:4874–4881

Kawahara Y, Ito K, Sun H, Aizawa H, Kanazawa I, Kwak S (2004) Glutamate receptors: RNA editing and death of motor neurons. Nature 427:801

Kawahara Y, Zinshteyn B, Sethupathy P, Iizasa H, Hatzigeorgiou AG, Nishikura K (2007a) Redirection of silencing targets by adenosine-to-inosine editing of miRNAs. Science 315:1137–1140

Kawahara Y, Zinshteyn B, Chendrimada TP, Shiekhattar R, Nishikura K (2007b) RNA editing of the microRNA-151 precursor blocks cleavage by the Dicer-TRBP complex. EMBO Rep 8:763–769

Keegan LP, Brindle J, Gallo A, Leroy A, Reenan RA, O'Connell MA (2005) Tuning of RNA editing by ADAR is required in *Drosophila*. EMBO J 24:2183–2193

Kim U, Wang Y, Sanford T, Zeng Y, Nishikura K (1994) Molecular cloning of cDNA for double-stranded RNA adenosine deaminase, a candidate enzyme for nuclear RNA editing. Proc Natl Acad Sci USA 91:11457–11461

Kim YG, Lowenhaupt K, Maas S, Herbert A, Schwartz T, Rich A (2000) The Zab domain of the human RNA editing enzyme ADAR1 recognizes Z-DNA when surrounded by B-DNA. J Biol Chem 275:26828–26833

Kim DD, Kim TT, Walsh T, Kobayashi Y, Matise TC, Buyske S, Gabriel A (2004) Widespread RNA editing of embedded Alu elements in the human transcriptome. Genome Res 14:1719–1725

Klaue Y, Källman AM, Bonin M, Nellen W, Öhman M (2003) Biochemical analysis and scanning force microscopy reveal productive and nonproductive ADAR2 binding to RNA substrates. RNA 9:839–846

Knight SW, Bass BL (2002) The role of RNA editing by ADARs in RNAi. Mol Cell 10:809–817

Kornblihtt AR, de la Mata M, Fededa JP, Munoz MJ, Nogues G (2004) Multiple links between transcription and splicing. RNA 10:1489–1498

Kumar M, Carmichael GG (1997) Nuclear antisense RNA induces extensive adenosine modifications and nuclear retention of target transcripts. Proc Natl Acad Sci USA 94:3542–3547

Kuo MY, Chao M, Taylor J (1989) Initiation of replication of the human hepatitis delta virus genome from cloned DNA: role of delta antigen. J Virol 63:1945–1950

Laurencikiene J, Källman AM, Fong N, Bentley DL, Öhman M (2006) RNA editing and alternative splicing: the importance of co-transcriptional coordination. EMBO Rep 7:303–307

Lehmann KA, Bass BL (1999) The importance of internal loops within RNA substrates of ADAR1. J Mol Biol 291:1–13

Lehmann KA, Bass BL (2000) Double-stranded RNA adenosine deaminases ADAR1 and ADAR2 have overlapping specificities. Biochemistry 39:12875–12884

Lejeune F, Maquat LE (2005) Mechanistic links between nonsense-mediated mRNA decay and pre-mRNA splicing in mammalian cells. Curr Opin Cell Biol 17:309–315

Levanon EY, Eisenberg E, Yelin R, Nemzer S, Hallegger M, Shemesh R, Fligelman ZY, Shoshan A, Pollock SR, Sztybel D, Olshansky M, Rechavi G, Jantsch MF (2004) Systematic identification of abundant A-to-I editing sites in the human transcriptome. Nat Biotechnol 22:1001–1005

Levanon EY, Hallegger M, Kinar Y, Shemesh R, Djinovic-Carugo K, Rechavi G, Jantsch MF, Eisenberg E (2005) Evolutionarily conserved human targets of adenosine to inosine RNA editing. Nucleic Acids Res 33:1162–1168

Lim LP, Lau NC, Garrett-Engele P, Grimson A, Schelter JM, Castle J, Bartel DP, Linsley PS, Johnson JM (2005) Microarray analysis shows that some microRNAs downregulate large numbers of target mRNAs. Nature 433:769–773

Liu Y, George CX, Patterson JB, Samuel CE (1997) Functionally distinct double-stranded RNA-binding domains associated with alternative splice site variants of the interferon-inducible double-stranded RNA-specific adenosine deaminase. J Biol Chem 272:4419–4428

Lomeli H, Mosbacher J, Melcher T, Hoger T, Geiger JR, Kuner T, Monyer H, Higuchi M, Bach A, Seeburg PH (1994) Control of kinetic properties of AMPA receptor channels by nuclear RNA editing. Science 266:1709–1713

Luciano DJ, Mirsky H, Vendetti NJ, Maas S (2004) RNA editing of a miRNA precursor. RNA 10:1174–1177

Luo GX, Chao M, Hsieh SY, Sureau C, Nishikura K, Taylor J (1990) A specific base transition occurs on replicating hepatitis delta virus RNA. J Virol 64:1021–1027

Maas S, Melcher T, Herb A, Seeburg PH, Keller W, Krause S, Higuchi M, O'Connell MA (1996) Structural requirements for RNA editing in glutamate receptor pre-mRNAs by recombinant double-stranded RNA adenosine deaminase. J Biol Chem 271:12221–12226

Maas S, Patt S, Schrey M, Rich A (2001) Underediting of glutamate receptor GluR-B mRNA in malignant gliomas. Proc Natl Acad Sci USA 98:14687–14692

Macbeth MR, Schubert HL, Vandemark AP, Lingam AT, Hill CP, Bass BL (2005) Inositol hexakisphosphate is bound in the ADAR2 core and required for RNA editing. Science 309:1534–1539

Melcher T, Maas S, Herb A, Sprengel R, Higuchi M, Seeburg PH (1996a) RED2, a brain-specific member of the RNA-specific adenosine deaminase family. J Biol Chem 271:31795–31798

Melcher T, Maas S, Herb A, Sprengel R, Seeburg PH, Higuchi M (1996b) A mammalian RNA editing enzyme. Nature 379:460–464

Mendell JT, Sharifi NA, Meyers JL, Martinez-Murillo F, Dietz HC (2004) Nonsense surveillance regulates expression of diverse classes of mammalian transcripts and mutes genomic noise. Nat Genet 36:1073–1078

Morse DP, Bass BL (1999) Long RNA hairpins that contain inosine are present in *Caenorhabditis elegans* poly(A)+ RNA. Proc Natl Acad Sci USA 96:6048–6053

Morse DP, Aruscavage PJ, Bass BL (2002) RNA hairpins in noncoding regions of human brain and *Caenorhabditis elegans* mRNA are edited by adenosine deaminases that act on RNA. Proc Natl Acad Sci USA 99:7906–7911

Mortillaro MJ, Blencowe BJ, Wei X, Nakayasu H, Du L, Warren SL, Sharp PA, Berezney R (1996) A hyperphosphorylated form of the large subunit of RNA polymerase II is associated with splicing complexes and the nuclear matrix. Proc Natl Acad Sci USA 93:8253–8257

Neugebauer KM, Roth MB (1997) Distribution of pre-mRNA splicing factors at sites of RNA polymerase II transcription. Genes Dev 11:1148–1159

Nishikura K (2004) Editing the message from A to I. Nat Biotechnol 22:962–963

Niswender CM, Copeland SC, Herrick-Davis K, Emeson RB, Sanders-Bush E (1999) RNA editing of the human serotonin 5-hydroxytryptamine 2C receptor silences constitutive activity. J Biol Chem 274:9472–9478

O'Connell MA, Krause S, Higuchi M, Hsuan JJ, Totty NF, Jenny A, Keller W (1995) Cloning of cDNAs encoding mammalian double-stranded RNA-specific adenosine deaminase. Mol Cell Biol 15:1389–1397

Ohlson J, Pedersen JS, Haussler D, Öhman M (2007) Editing modifies the GABA(A) receptor subunit alpha3. RNA 13:698–703

Öhman M, Källman AM, Bass BL (2000) In vitro analysis of the binding of ADAR2 to the pre-mRNA encoding the GluR-B R/G site. RNA 6:687–697

Palladino MJ, Keegan LP, O'Connell MA, Reenan RA (2000a) A-to-I pre-mRNA editing in *Drosophila* is primarily involved in adult nervous system function and integrity. Cell 102:437–449

Palladino MJ, Keegan LP, O'Connell MA, Reenan RA (2000b) dADAR, a *Drosophila* double-stranded RNA-specific adenosine deaminase is highly developmentally regulated and is itself a target for RNA editing [In Process Citation]. RNA 6:1004–1018

Patterson JB, Samuel CE (1995) Expression and regulation by interferon of a double-stranded-RNA-specific adenosine deaminase from human cells: evidence for two forms of the deaminase. Mol Cell Biol 15:5376–5388

Patton DE, Silva T, Bezanilla F (1997) RNA editing generates a diverse array of transcripts encoding squid Kv2 K+ channels with altered functional properties. Neuron 19:711–722

Pinto Desterro JM, Keegan LP, Jaffray E, Hay RT, O'Connell MA, Carmo-Fonseca M (2005) SUMO-1 modification alters ADAR1 editing activity. Mol Biol Cell (in press)

Polson AG, Bass BL (1994) Preferential selection of adenosines for modification by double-stranded RNA adenosine deaminase. EMBO J 13:5701–5711

Poulsen H, Nilsson J, Damgaard CK, Egebjerg J, Kjems J (2001) CRM1 mediates the export of ADAR1 through a nuclear export signal within the Z-DNA binding domain. Mol Cell Biol 21:7862–7871

Prasanth KV, Prasanth SG, Xuan Z, Hearn S, Freier SM, Bennett CF, Zhang MQ, Spector DL (2005) Regulating gene expression through RNA nuclear retention. Cell 123:249–263

Ramos A, Grunert S, Adams J, Micklem DR, Proctor MR, Freund S, Bycroft M, St Johnston D, Varani G (2000) RNA recognition by a Staufen double-stranded RNA-binding domain. EMBO J 19:997–1009

Reenan RA (2005) Molecular determinants and guided evolution of species-specific RNA editing. Nature 434:409–413

Rosenthal JJ, Bezanilla F (2002) Extensive editing of mRNAs for the squid delayed rectifier K+ channel regulates subunit tetramerization. Neuron 34:743–757

Rueter SM, Dawson TR, Emeson RB (1999) Regulation of alternative splicing by RNA editing. Nature 399:75–80

Ryman K, Fong N, Bratt E, Bentley DL, Öhman M (2007) The C-terminal domain of RNA pol II helps ensure that editing precedes splicing of the GluR-B transcript. RNA 35:3723–3732

Ryter JM, Schultz SC (1998) Molecular basis of double-stranded RNA-protein interactions: structure of a dsRNA-binding domain complexed with dsRNA. EMBO J 17:7505–7513

Sallacz NB, Jantsch MF (2005) Chromosomal storage of the RNA-editing enzyme ADAR1 in *Xenopus* oocytes. Mol Biol Cell 16:3377–3386

Sansam CL, Wells KS, Emeson RB (2003) Modulation of RNA editing by functional nucleolar sequestration of ADAR2. Proc Natl Acad Sci USA 100:14018–14023

Scadden AD (2005) The RISC subunit Tudor-SN binds to hyper-edited double-stranded RNA and promotes its cleavage. Nat Struct Mol Biol 12:489–496

Scadden AD, Smith CW (1997) A ribonuclease specific for inosine-containing RNA: a potential role in antiviral defence? EMBO J 16:2140–2149

Scadden AD, Smith CW (2001a) RNAi is antagonized by A–>I hyper-editing. EMBO Rep 2:1107–1111

Scadden AD, Smith CW (2001b) Specific cleavage of hyper-edited dsRNAs. EMBO J 20:4243–4252

Schoft VK, Schopoff S, Jantsch MF (2007) Regulation of glutamate receptor B pre-mRNA splicing by RNA editing. Nucleic Acids Res 35:3723–3732

Seeburg PH, Hartner J (2003) Regulation of ion channel/neurotransmitter receptor function by RNA editing. Curr Opin Neurobiol 13:279–283

Seeburg PH, Higuchi M, Sprengel R (1998) RNA editing of brain glutamate receptor channels: mechanism and physiology. Brain Res Brain Res Rev 26:217–229

Sommer B, Kohler M, Sprengel R, Seeburg PH (1991) RNA editing in brain controls a determinant of ion flow in glutamate-gated channels. Cell 67:11–19

Stefl R, Allain FH (2005) A novel RNA pentaloop fold involved in targeting ADAR2. RNA 11:592–597

Stefl R, Xu M, Skrisovska L, Emeson RB, Allain FH (2006) Structure and specific RNA binding of ADAR2 double-stranded RNA binding motifs. Structure 14:345–355

Stephens OM, Yi-Brunozzi HY, Beal PA (2000) Analysis of the RNA-editing reaction of ADAR2 with structural and fluorescent analogues of the GluR-B R/G editing site. Biochemistry 39:12243–12251

Stephens OM, Haudenschild BL, Beal PA (2004) The binding selectivity of ADAR2's dsRBMs contributes to RNA-editing selectivity. Chem Biol 11:1239–1250

Strehblow A, Hallegger M, Jantsch MF (2002) Nucleocytoplasmic distribution of human RNA-editing enzyme ADAR1 is modulated by double-stranded RNA-binding domains, a leucine-rich export signal, and a putative dimerization domain. Mol Biol Cell 13:3822–3835

Tanoue A, Koshimizu TA, Tsuchiya M, Ishii K, Osawa M, Saeki M, Tsujimoto G (2002) Two novel transcripts for human endothelin B receptor produced by RNA editing/alternative splicing from a single gene. J Biol Chem 277:33205–33212

Tonkin LA, Bass BL (2003) Mutations in RNAi rescue aberrant chemotaxis of ADAR mutants. Science 302:1725

Tonkin LA, Saccomanno L, Morse DP, Brodigan T, Krause M, Bass BL (2002) RNA editing by ADARs is important for normal behavior in *Caenorhabditis elegans*. EMBO J 21:6025–6035

Vitali P, Basyuk E, Le Meur E, Bertrand E, Muscatelli F, Cavaille J, Huttenhofer A (2005) ADAR2-mediated editing of RNA substrates in the nucleolus is inhibited by C/D small nucleolar RNAs. J Cell Biol 169:745–753

Wagner RW, Yoo C, Wrabetz L, Kamholz J, Buchhalter J, Hassan NF, Khalili K, Kim SU, Perussia B, McMorris FA et al. (1990) Double-stranded RNA unwinding and modifying activity is detected ubiquitously in primary tissues and cell lines. Mol Cell Biol 10:5586–5590

Wang Q, O'Brien PJ, Chen CX, Cho DS, Murray JM, Nishikura K (2000) Altered G protein-coupling functions of RNA editing isoform and splicing variant serotonin2C receptors. J Neurochem 74:1290–1300

Wang Q, Miyakoda M, Yang W, Khillan J, Stachura DL, Weiss MJ, Nishikura K (2004) Stress-induced apoptosis associated with null mutation of ADAR1 RNA editing deaminase gene. J Biol Chem 279:4952–4961

Wang Q, Zhang Z, Blackwell K, Carmichael GG (2005) Vigilins bind to promiscuously A-to-I-edited RNAs and are involved in the formation of heterochromatin. Curr Biol 15:384–391

Wong SK, Sato S, Lazinski DW (2001) Substrate recognition by ADAR1 and ADAR2. RNA 7:846–858

Wu H, Henras A, Chanfreau G, Feigon J (2004) Structural basis for recognition of the AGNN tetraloop RNA fold by the double-stranded RNA-binding domain of Rnt1p RNase III. Proc Natl Acad Sci USA 101:8307–8312

Xia S, Yang J, Su Y, Qian J, Ma E, Haddad GG (2005) Identification of new targets of *Drosophila* pre-mRNA adenosine deaminase. Physiol Genomics 20:195–202

Yang JH, Nie Y, Zhao Q, Su Y, Pypaert M, Su H, Rabinovici R (2003) Intracellular localization of differentially regulated RNA-specific adenosine deaminase isoforms in inflammation. J Biol Chem 278:45833–45842

Yang W, Wang Q, Kanes SJ, Murray JM, Nishikura K (2004) Altered RNA editing of serotonin 5-HT2C receptor induced by interferon: implications for depression associated with cytokine therapy. Brain Res Mol Brain Res 124:70–78

Yang W, Wang Q, Howell KL, Lee JT, Cho DS, Murray JM, Nishikura K (2005) ADAR1 RNA deaminase limits short interfering RNA efficacy in mammalian cells. J Biol Chem 280:3946–3953

Yang W, Chendrimada TP, Wang Q, Higuchi M, Seeburg PH, Shiekhattar R, Nishikura K (2006) Modulation of microRNA processing and expression through RNA editing by ADAR deaminases. Nat Struct Mol Biol 13:13–21

Yi-Brunozzi HY, Stephens OM, Beal PA (2001) Conformational changes that occur during an RNA-editing adenosine deamination reaction. J Biol Chem 276:37827–37833

Zhang Z, Carmichael GG (2001) The fate of dsRNA in the nucleus: a p54(nrb)-containing complex mediates the nuclear retention of promiscuously A-to-I edited RNAs. Cell 106:465–475

Zorio DA, Bentley DL (2004) The link between mRNA processing and transcription: communication works both ways. Exp Cell Res 296:91–97

Insertion/Deletion Editing in *Physarum polycephalum**

Jonatha M. Gott (✉) and Amy C. Rhee

> *"Every revolution evaporates and leaves behind only the slime*
> *of a new bureaucracy."*
>
> Franz Kafka

Abstract Mitochondrial gene expression in the acellular slime mold *Physarum polycephalum* requires a diverse array of RNA editing events. Virtually all transcripts encoded in the organelle are subject to changes at the RNA level, including

---

Center for RNA Molecular Biology, 10900 Euclid Avenue, Case Western Reserve University, Cleveland, OH 44106, USA; jmg13@case.edu

* After submission of this chapter, the cloning and characterization of the *Physarum* mitochondrial RNA polymerase was published by Dennis Miller and colleagues (Miller et al. 2006).

both mRNAs and stable RNAs. Roughly 500 editing events involving nucleotide insertion or deletion have been confirmed thus far; these occur co-transcriptionally. Base changes also occur in *Physarum* mitochondria, but these are much rarer and occur post-transcriptionally. This chapter focuses on the experimental approaches used to dissect the unique mechanism of insertion/deletion editing in *Physarum* mitochondria.

1 Introduction

The mitochondrial genome of *Physarum polycephalum* lacks open reading frames (ORFs) that encode conventional mitochondrial proteins. Despite this apparent lack of genes, a normal complement of mitochondrial messenger RNAs (mRNAs) is present. Creation of these ORFs depends upon RNA editing within the organelle (Table 1 and references therein). Editing is not limited to mRNAs in *Physarum* mitochondria; stable RNAs, including the mitochondrially encoded transfer RNAs (tRNAs; Antes et al. 1998), the large and small ribosomal RNAs (rRNAs; Miller et al. 1993; Mahendren et al. 1994), and a newly discovered mitochondrial 5S

Table 1 Characterized editing events in *Physarum* mitochondria

RNA	Nucleotide insertion/base substitution	Reference
*atp*1 mRNA	54 +C	Mahendran et al. (1991)
*atp*8 mRNA	9 +C	Gott et al. (2005)
*atp*9 mRNA	9 +C	Miller et al. (1993)
cytb mRNA	31 +C, 6 +U, 2 +CU, 1 +GC	Wang et al. (1999)
*cox*1 mRNA	59 +C, 1 +U, 1 +UA, 1 +CU, 1 +GU, four C to U changes	Gott et al. (1993); Visomirski-Robic and Gott (1995)
*cox*2 mRNA	33 +C	Gott et al. (2005)
*nad*1 mRNA	38 +C, 2 +U	Unpublished data in Miller et al. (1993)
*nad*2 mRNA	55 +C, 4 +U, 1 −AAA	Gott et al. (2005)
*nad*4L mRNA	13 +C	Gott et al. (2005)
*nad*6 mRNA	18 +C, 1 +U	Gott et al. (2005)
*nad*7 mRNA	37 +C, 4 +U, 2 +CU, 2 +UU, 1 +GU	Takano et al. (2001)
Small rRNA	40 +C, 2 +U, 1 +CU, 2 +AA	Mahendran et al. (1994)
Large rRNA	52 +C, 2 +CU, 1 +GU, 2 +AA	Unpublished data in Miller et al. (1993)
5S rRNA	2 +C	Gott et al. (unpublished data)
tRNA[K]	1 +C	Antes et al. (1998)
tRNA[E]	1 +C, 1 +U	Antes et al. (1998)
tRNA[P]	2 +C	Antes et al. (1998)
tRNA[M1]	2 +C, one U to G change	Antes et al. (1998); Gott et al. (unpublished data)
tRNA[M2]	One C to G change	Gott et al. (unpublished data)

rRNA candidate (J.M. Gott, B. Somerlot, O. Kourennaia, and M. Gray, unpublished data), are also edited. Thus, all aspects of mitochondrial gene expression in *Physarum* are dependent upon accurate RNA editing.

2 Types of Editing in the Mitochondria of *Physarum polycephalum*

At least four distinct types of editing occur within the mitochondria of *Physarum polycephalum* (Table 1). The vast majority of editing events involve the insertion of non-encoded nucleotides. Of the nearly 500 nucleotide insertion sites characterized thus far, 90% involve the addition of single C residues at specific sites, with the remaining instances of nucleotide insertions involving single U and a subset of dinucleotide insertions (UU, AA, CU, GU, GC, and UA; see Table 1 and references therein). The other types of editing that occur in *Physarum* mitochondria are much less abundant. Only a single site of nucleotide deletion has been reported thus far, which involves the deletion of three adjacent encoded A residues within the *nad*2 mRNA (Gott et al. 2005). Similarly, only four instances of C to U changes have been observed, all within the *cox*1 mRNA (Gott et al. 1993). Finally, two of the mitochondrially encoded tRNAs are subject to editing at their 5′ ends (J.M. Gott, B. Somerlot, O. Kourennaia, and M. Gray, unpublished data) in a manner reminiscent of tRNA editing in the mitochondria of *Acanthamoeba castellanii* (Lonergan and Gray 1993), and chytridiomycete fungi (Laforest et al. 1997). This unprecedented diversity of editing types within a single organelle makes *Physarum* a unique system in which to study RNA editing.

3 Functions of Editing in *Physarum* Mitochondria

Because the *Physarum* mitochondrial genome is AT-rich (~75% A+T), stop codons occur frequently in all three reading frames of potential genes. In the absence of editing, truncated versions of the mitochondrial proteins would be made, often with long stretches of amino acids that diverge considerably from consensus sequences derived from the generally well-conserved mitochondrial proteins. An example is shown in Fig. 1A, which compares the predicted translational readouts of the *atp*8 gene (mtDNA) with that from the mRNA sequence (cDNA). Because the first editing event falls within the second codon, the predicted protein products from the gene and mRNA diverge immediately. If UAG codons are used as stop codons in *Physarum* mitochondria, the predicted polypeptide encoded from the *atp*8 gene would only be nine amino acids long. However, given that all characterized mRNAs from *Physarum* mitochondria have a UAA stop codon, UAG may be used as a sense codon in this system. If the first

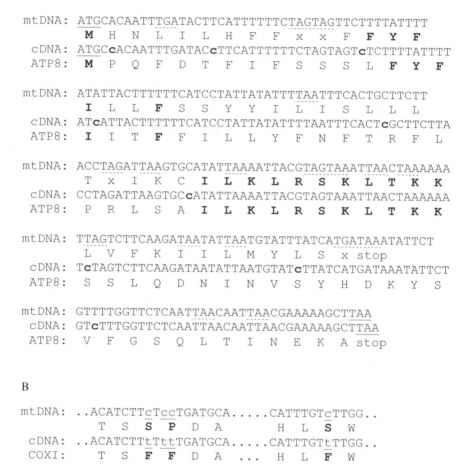

Fig. 1 Comparison of predicted translation products from genes and mRNAs from *Physarum* mitochondria. **A** Comparison of the *atp*8 gene (mtDNA) and mRNA (cDNA). The actual start and stop codons are *underlined*, with potential stop codons in the mtDNA indicated with a *dotted underline*. Amino acids that are common to both predicted products are shown *in bold*. Inserted C residues are each shown by a *bold lowercase* c in the cDNA sequence. **B** Comparison of the pertinent regions of the *cox*1 gene (mtDNA) and mRNA (cDNA). Sites of C to U changes in the mRNA are each indicated by an *underlined lowercase* t in the cDNA. Amino acids that are predicted to be altered due to these C to U changes are shown in *bold*. See text for details

in-frame UAA were used to terminate translation of the genomic version (mtDNA), then the sequence of the protein would still be significantly shorter (57 amino acids) than that encoded by the mRNA (76 amino acids), and would bear little resemblance to other *atp*8 gene products (Fig. 1A). In contrast, the insertion of "extra" nucleotides creates multiple frameshifts within the *atp*8 mRNA, resulting in the formation of an ORF predicted to yield a polypeptide highly homologous to those found in other species (Fig. 1A; Gott et al. 2005).

The *cox*1 mRNA, which contains single C insertions, a single U insertion, and three dinucleotide insertions, also contains four C to U changes that result in changes at the amino acid level (Table 1, Fig. 1B). These editing events do not appear to be developmentally regulated, as all single C, single U, and C to U changes examined in the *cox*1 mRNA are virtually completely edited at all lifecycle stages (Rundquist and Gott 1995). There is also no indication that editing of other mitochondrial RNAs is developmentally regulated, but this has not been systematically investigated.

Insertional editing events in stable RNAs (rRNAs, 5S RNA, and tRNAs) lead to the generation of primary, secondary, and tertiary structural elements that are highly conserved and/or functionally important in other organisms. Examples include the highly conserved TΨC sequence of tRNAglu, which requires a U addition prior to nucleotide modification (Antes et al. 1998), and the addition of two A residues within the small subunit rRNA (Mahendran et al. 1994) that, in *E. coli*, are involved in the proofreading of tRNAs in the A site during translation (*E coli* numbering A1492 and A1493; Yoshizawa et al. 1999). Two of the five mitochondrially encoded tRNAs are also subject to editing at their 5′ ends. In this case, editing corrects apparent mismatches present at the top of the acceptor stem of tRNAmet1 and tRNAmet2 that are presumably not compatible with tRNA function (J.M. Gott, B. Somerlot, O. Kourennaia, and M. Gray, unpublished data).

4   Ramifications of Editing for Gene Identification in the Mitochondria of *Physarum polycephalum*

The sequence of the entire mitochondrial genome of *Physarum polycephalum* was published by Takano et al. (2001). However, because insertional editing events in *Physarum* mitochondria are extremely frequent, averaging one insertion for every 25 nucleotides in mRNAs, and one insertion for every 40 nucleotides in stable RNAs (Miller et al. 1993), standard "gene finding" algorithms fail to identify many of the genes of which the transcripts require editing to create a functional RNA. To overcome this problem, Ralf Bundschuh and colleagues have developed algorithms that account for the characteristics of *Physarum* editing sites (Bundschuh 2004; R. Bundschuh, H-Y. Lee, and T. Liu, unpublished data). These algorithms have proven highly effective at both finding mitochondrial genes and predicting sites of nucleotide insertion, particularly when probabilistic predictions are included in the algorithm (Gott et al. 2005). Recent improvements to the original algorithm have led to the identification of at least 15 additional potential genes of which the mRNAs have not yet been characterized (R. Bundschuh, C. Ainsley, H-Y. Lee, and T. Liu, unpublished data).

5   Mechanisms of Editing in *Physarum* Mitochondria

5.1   Evidence for Multiple Editing Pathways

Based on available evidence, it is likely that there are at least three distinct editing pathways in the mitochondria of *Physarum polycephalum*. The insertion of non-encoded nucleotides as single or dinucleotides is a co-transcriptional process, as described in more detail below. The mechanism of nucleotide deletion has not yet been investigated, but we hypothesize that this may also be a co-transcriptional process related to nucleotide insertion, based on both the characterization of steady-state RNAs made in vivo and misediting events observed in vitro. Unlike what is observed in other organisms in which RNA editing occurs, which generally contain a mixture of edited, unedited, and partially edited molecules, RNAs present in *Physarum* mitochondria are virtually 100% edited at all nucleotide insertion sites (Rundquist and Gott 1995) as well as at the AAA deletion site (Gott et al. 2005), arguing that both forms of editing must be tightly coupled to transcription. In vitro, misediting occurs at a low level in *Physarum* RNAs during run-on transcription using preformed mitochondrial transcription elongation complexes (Byrne et al. 2002). One form of misediting that was observed in these experiments is the deletion of three encoded nucleotides immediately adjacent to a C insertion site (Byrne et al. 2002). Thus, at least in vitro, nucleotide deletions can arise during RNA synthesis.

Other forms of editing that occur in *Physarum* mitochondria, including both the C to U changes and the correction of mismatches at the 5′ end of tRNAs, are most likely distinct post-transcriptional processes. *Nascent cox*1 mRNAs made both in vivo (Gott and Visomirski-Robic 1998; Byrne and Gott 2004) and in vitro (Visomirski-Robic and Gott 1995; Byrne and Gott 2004) have Cs, rather than Us, at the sites of C to U substitutions, despite the fact that *mature* mitochondrial RNAs contain almost exclusively Us at these positions (Fig. 2A; Gott et al. 1993). Mechanistic studies of these C to U substitutions have not been carried out as yet, but it is likely that the C to U changes are site-specific deamination events similar to those observed in the mammalian *apo*B mRNA (Johnson et al. 1993) and plant mitochondria (Rajasekhar and Mulligan 1993). Likewise, *Physarum* mitochondrial tRNAmet1 and tRNAmet2 *precursors* contain the nucleotides encoded in the genome, rather than the Gs present at their 5′ ends in the mature species (Fig. 2B), suggesting that the U to G (in tRNAmet1) and C to G (in tRNAmet2) changes observed at the 5′ end of these tRNAs occur after processing of the nascent transcripts (J.M. Gott, B. Somerlot, O. Kourennaia, and M. Gray, unpublished data). Because neither the C to U changes nor tRNA 5′ editing mechanisms have been investigated as yet in *Physarum* mitochondria, only experiments relating to the mechanism of insertion editing are discussed in more detail below.

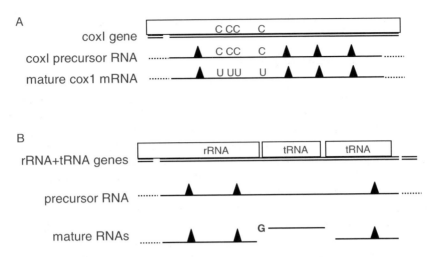

Fig. 2 Schematic illustration of the timing of various editing events in *Physarum* mitochondria. **A** Editing events within the *cox*1 precursor and mature mRNA are indicated. *Triangles* indicate the presence of "extra" nucleotides added via insertional editing; the presence of either a C or a U at C to U sites within the RNA is indicated by *letters*. **B** Editing events within the ribosomal RNA/tRNA precursor and mature RNAs are indicated. *Triangles* indicate the presence of "extra" nucleotides added via insertional editing; G indicates the replacement of the tRNA 5′ nucleotide

5.2   Co-Transcriptional Nucleotide Insertions (and Potentially Deletions)

Initial characterization of steady-state RNA pools in *Physarum* mitochondria suggested that the mechanism of nucleotide insertion in *Physarum* differs significantly from U insertion/deletion editing in the kinetoplasts of trypanosomatids. Unlike what is observed in trypanosomes, where a range of potential editing intermediates were observed, essentially all mitochondrial RNAs in *Physarum* are fully edited throughout their entire length. Based on this finding, we initially hypothesized that *Physarum* editing occurs in a 5′ to 3′ direction on nascent RNAs, in a process that is closely coupled to transcription (Gott et al. 1993). While both of these predictions have been validated (Visomirski-Robic and Gott 1997a, b), in vitro studies in coupled transcription/editing systems derived from *Physarum* mitochondria led to the surprising conclusion that editing actually occurs at the very 3′ end of nascent mitochondrial transcripts (Cheng et al. 2001). Superficially, this finding suggested that *Physarum* editing might occur via a polymerase stuttering mechanism similar to that observed in paramyxoviruses (Vidal et al. 1990). However, the context of editing sites, and the effects of altering nucleotide concentrations during in vitro transcription are quite different in the two systems, arguing for separate mechanisms (Cheng et al. 2001). Thus, the mechanism of nucleotide insertion in *Physarum* is distinct from other characterized forms of insertional RNA editing.

The conclusion that the insertion of single C residues occurs during transcription is based upon a number of lines of evidence. First, labeled RNAs made in isolated mitochondria are efficiently edited under conditions of limited RNA synthesis (Visomirski-Robic and Gott 1995). Second, editing occurs within 14 nucleotides of the 3′ end of nascent transcript (Visomirski-Robic and Gott 1997b). Third, once unedited RNA is synthesized, it cannot be chased into edited RNA under conditions in which newly synthesized RNAs are edited, i.e., there is a limited window in which editing can occur (Visomirski-Robic and Gott 1997a). Fourth, editing occurs with a 5′ to 3′ polarity (Visomirski-Robic and Gott 1997a). Fifth, in incompletely edited RNAs synthesized by crude transcription/editing complexes in vitro, the edited and unedited sites are interspersed, with different patterns observed in different molecules (Byrne and Gott 2002). Finally, the efficiency of C insertion at a given editing site is inversely proportional to the concentration of the encoded nucleotide immediately downstream of the insertion site (Cheng et al. 2001), suggesting that editing and transcription elongation are competing processes. Taken together, these data are most easily explained by a model for editing in which the non-encoded nucleotides are added at the 3′ end of the nascent RNA during synthesis.

Insertion of single Us and dinucleotides almost certainly occurs via the same basic mechanism as for C insertions, although separate or additional factors may be required at these sites. This is based on experiments involving labeled transcripts made in isolated mitochondria and crude mitochondrial transcription complexes (Visomirski-Robic and Gott 1997a; J.M. Gott and Y-W. Cheng, unpublished data), S1 nuclease protection experiments (Visomirski-Robic and Gott 1997b), and analysis of run-on transcripts made from chimeric templates (Byrne et al. 2002; Byrne and Gott 2004). Similarly, although the timing of nucleotide deletion has not yet been examined, it is likely that deletion of encoded nucleotides is also a co-transcriptional process. This prediction is based on (1) the analysis of total RNA made in vivo, where essentially all RNA contains the deletion (Gott et al. 2005) but not the post-transcriptional C to U changes (Gott and Visomirski-Robic 1998), (2) the presence of the deletion in all individual cDNA clones isolated (Gott et al. 2005), and (3) the presence of occasional misediting at C insertion sites involving three nucleotide deletions generated during run-on transcription in vitro (Byrne et al. 2002).

We hypothesize that a number of steps are required for the correct insertion of non-encoded nucleotides (Fig. 3). Since all insertion events observed thus far occur within the coding region of mRNAs, or the portions of stable RNA precursors that are found in the mature RNAs, nucleotide insertion must occur in the context of transcription elongation by the *Physarum* mitochondrial RNA polymerase (mtRNAP). Upon reaching an editing site, the polymerase has to stop normal template-directed transcription to allow the 3′ end of the RNA to enter the active site of the editing machinery. This is likely accomplished via conformational

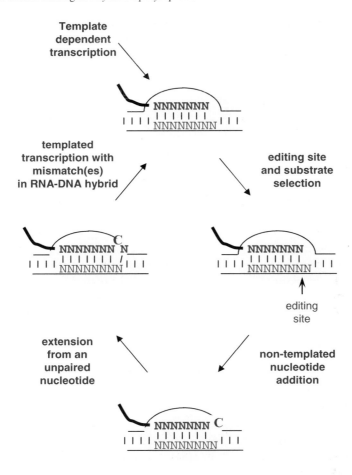

Fig. 3 Hypothetical transcription/editing cycle. The mitochondrial DNA is shown with *thin lines* and *standard text*. Nascent RNAs are indicated by *thick lines* and *bold text*, with non-templated Cs shown in *outline form*. Base pairing is indicated by *vertical lines*. The *Physarum* mitochondrial RNA polymerase is depicted as a *shaded oval*. Proposed steps in the transcription/editing cycle are shown

changes in the polymerase, RNA–DNA hybrid, or both. How recognition of an editing site is achieved is currently not known, nor is it clear how the specific nucleotide or nucleotides to be added at a particular site are identified (discussed below). Once the non-templated nucleotide has been incorporated at the 3′ end of the RNA, this *unpaired* nucleotide must be extended in a template-directed manner, presumably by the *Physarum* mtRNAP. Finally, "normal" transcription in which nucleotides are again added to a base-paired nucleotide in a template-dependent fashion must resume. However, in this case the mtRNAP must accommodate an "extra", unpaired nucleotide in the RNA–DNA hybrid. This may be particularly problematic in instances of dinucleotide insertions, where the RNA–DNA hybrid would contain two extra nucleotides, or potentially, the triple nucleotide deletion, in which case three nucleotides in the DNA template would not be paired. It seems likely that additional factors are required to stabilize the transcription elongation complex at these sites, as discussed in more detail below.

6 Requirements for Nucleotide Insertion

6.1 Cis-Acting Elements

Because editing and transcription are coupled, cis-acting elements involved in editing could be present at the level of DNA, RNA, or both. Removal of most of the upstream RNA using oligonucleotide-directed RNase H cleavage prior to run-on transcription/editing does not change the efficiency of editing (J.M. Gott and A. Majewski, unpublished data), indicating that if RNA sequences are involved in nucleotide insertion, then the essential region falls within the last 10–15 nucleotides of the RNA. These data also argue against extensive backtracking by the mtRNAP being required for RNA editing. The significance of this finding is discussed below.

The DNA sequences required for nucleotide insertion localize to the region immediately surrounding individual editing sites. This conclusion is based on experiments involving both run-off transcription and chimeric templates. When the mitochondrial DNA (mtDNA) in transcription elongation complexes (mtTECs) is digested by a restriction endonuclease that cleaves 14 base pairs downstream of an editing site, the extent of editing at that site is similar to the level observed in untreated mtTECs, demonstrating that only a limited amount of downstream DNA is required for editing site recognition and utilization (J.M. Gott and A. Majewski, unpublished data). If all but six base pairs of downstream DNA are removed, however, editing is not observed, indicating that the region between 6–14 bp contains information necessary for RNA editing and/or the stability of the transcription/editing complex (J.M. Gott and A. Majewski, unpublished data). To distinguish between these possibilities, a series of chimeric templates have been made in which downstream DNA is substituted by other regions of mitochondrial DNA or exogenous DNA sequences (Fig. 4; Byrne and Gott 2002; A.C. Rhee, unpublished data). These experiments indicate that 11 bp of native downstream DNA is sufficient to support efficient nucleotide insertion (A.C. Rhee, unpublished data); further definition of the downstream DNA requirements is in progress. A similar strategy has been used to define the upstream DNA requirements (Fig. 4). Using circular templates generated by restriction enzyme cleavage and self-ligation of the mitochondrial DNA in mtTECs, the upstream boundary has been defined as 8–9 bp of native sequence (A.C. Rhee, unpublished data).

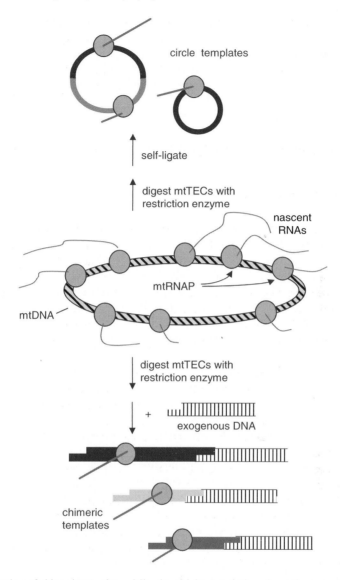

Fig. 4 Generation of chimeric templates. Mitochondrial transcription elongation complexes consisting of the mitochondrial genome (mtDNA), shown as a *diagonally hatched circle*, and associated proteins can be treated with restriction enzymes and DNA ligase to generate two types of templates. Subsequent ligation in the absence of exogenous DNA yields predominantly circular templates (*upper portion* of the figure), whereas inclusion of exogenous DNA (*vertically striped line*) generates chimeric templates (*lower portion* of the figure)

6.2 Trans-Acting Factors

Insertional editing requires one or more trans-acting factors, at least one of which appears to be associated with the DNA template. This conclusion is based on experiments involving chimeric templates generated by ligating exogenous DNA to fragmented mtTECs (Fig. 4; Byrne and Gott 2002). When mtDNA present in mtTECs is digested with a restriction enzyme, ligated to exogenous DNA, and used as the template for run-on transcription, the portion of the RNA that was generated from the "native" template is edited by nucleotide insertion, while the RNA resulting from transcription of "naked" DNA is not (Byrne and Gott 2002). The same result was seen irregardless of the source of DNA; PCR products, isolated plasmid DNA, and deproteinized mitochondrial DNA all failed to support nucleotide insertion in these chimeric template experiments (Byrne and Gott 2002). The latter result using deproteinized mitochondrial DNA was particularly intriguing, as it ruled out the possibility that some covalent DNA modification present in *Physarum* mitochondrial DNA was sufficient to direct RNA editing.

It is likely that separate or additional factors are needed, at least at a subset of insertion sites. This conclusion is based on patterns of editing and misediting observed in vitro. Unlike the RNA made in vivo, RNAs made during run-on transcription/editing in vitro using isolated mtTECs are not completely edited (Cheng and Gott 2000; Byrne et al. 2002). The extent of editing at a given site can vary from nearly 0 to almost 100%, depending on conditions, and is particularly sensitive to nucleotide concentrations (Cheng et al. 2001). Characterization of cDNAs derived from individual run-on transcripts uncovered a low level of misediting, with roughly 5–7% of the insertion sites affected in any given experiment (Byrne et al. 2002). Such misediting events include the addition of an "incorrect" nucleotide (or nucleotides) at an insertion site, the deletion of 1–3 encoded nucleotides immediately adjacent to an insertion site, and the generation of "intersite deletions", which lack all of the encoded nucleotides between two editing sites (Byrne et al. 2002). An intriguing finding from these types of experiments is that certain dinucleotide insertion sites are subject to systematic misediting in vitro, suggesting that additional factors are required at these sites (Byrne and Gott 2004). This observation is consistent with the findings of Horton and Landweber (2000), who determined that the patterns of editing within the *cox*1 mRNA varies in closely related organisms. Whereas RNAs from *Didymium nigripes* and *Stemonitis flavogenita* demonstrated a similar range and number of insertional editing events to that observed previously in *Physarum polycephalum* (which contains 59 C insertions, one U insertion, and three dinucleotide insertions within its *cox*1 mRNA), *cox*1 RNAs from *Arcyria cinerea* have limited numbers of both U and C insertions, but lack dinucleotide insertions, and those from *Clastoderma debaryanum* have only four U insertions (Horton and Landweber 2000). These results suggest that the ability to insert single uridines arose first, with the other insertion activities arising later, presumably upon the acquisition of additional factors (Horton and Landweber 2000).

7 Potential Roles for Editing Factors

There are a number of roles that factors required for editing might play. These include (1) the identification of editing sites, (2) specification of the nucleotide(s) to be added, (3) conformational changes required to place the 3' end of the nascent RNA in the editing active site, (4) addition of the "extra", uncoded nucleotide, (5) assistance with extension, in a template-directed fashion, from an unpaired nucleotide at the 3' end of the transcript (presumably by the mitochondrial RNA polymerase), and (6) stabilization of the transcription/editing complex, including any of the stages mentioned above and/ or the accommodation of an altered RNA–DNA hybrid. These possible roles are discussed briefly below.

7.1 Editing Site Recognition

The generation of precise intersite deletions during run-on transcription in vitro suggests that editing sites are marked in some way (Byrne et al. 2002), but it is still unclear how this is accomplished. Obvious candidates for such a signal include binding sites for specific proteins, and covalent modifications within the DNA at or near editing sites. However, no consistent pattern of protein footprints could be distinguished in the vicinity of editing sites in experiments using dimethyl sulfate or potassium permanganate (J.M. Gott and L. Tsujikawa, unpublished data), suggesting that identification of editing sites is not accomplished by binding of an editing factor to specific sites.

Another possible means of marking editing sites would be modified nucleotides within the DNA, but the low level of modifications that was observed upon mass spectrometric analysis of *Physarum* mtDNA would not be sufficient to account for the roughly 500 known insertional editing sites in the mitochondrial genome (P. Crain and J.M. Gott, unpublished data), making this an unlikely source of editing site specification. Other possible means of marking editing sites include intrinsic pause signals, which could include structural perturbations in the DNA (such as bends or Z-DNA formation), or binding of non-specific proteins that serve to slow the transcription rate of the polymerase, perhaps enhancing its ability to respond to editing signals. In this regard, it is important to note that this effect is specific for editing sites; increasing the dwell time of the *Physarum* mtRNAP at other sites does not lead to the insertion of non-encoded nucleotides (Cheng et al. 2001). Inversely, misediting events such as the insertion of non-encoded G residues occur only at nucleotide insertion sites; no such misinsertion events have been observed at non-editing sites, again arguing that insertion sites have unique characteristics (Byrne et al. 2002).

7.2   Editing Site Utilization

It is currently not known how the identity of the nucleotide(s) added at an editing site is specified, and it is likely that one or more factors may be required for this process. No potential nucleic acid "templates" analogous to the guide RNAs found in trypanosomes have been detected using a range of experimental approaches (unpublished data), but such negative results are obviously inconclusive. Another possibility is that non-Watson-Crick pairings are used to specify the inserted nucleotide. This strategy is illustrated by the Y family of DNA polymerases, which can use the Hoogsteen edge of a template base in the *syn* conformation (Nair et al. 2004). It is also possible that the "template" for nucleotide addition is derived from a combination of functional groups contributed by both protein and nucleic acid residues, as observed for CCA-adding enzymes (Yue et al. 1998; Xiong and Steitz 2004; Tomita et al. 2004). Protein side chains could be contributed by either the polymerase or trans-acting factors, while the nucleic acid groups could be contributed by the template or non-template strands of the DNA or the elongating RNA itself. Additional factors may also be involved in the switch between transcription elongation and editing and/or the resumption of templated transcription from an unpaired 3′ nucleotide.

Although other factors are almost certainly required for editing, the mitochondrial RNA polymerase itself is the leading candidate for the actual insertion activity. In addition to its role in normal transcription, it is likely that the *Physarum* mtRNAP is responsible for the templated addition of nucleotides from the unpaired nucleotide that results from editing. It must also stop templated transcription in response to signals demarcating an editing site, and accommodate an extra nucleotide (or nucleotides) in the RNA–DNA hybrid once templated transcription resumes. Since DNA oligonucleotides bound to the RNA immediately upstream of an insertion site have no effect on editing (J.M. Gott and A. Majewski, unpublished data), it appears that polymerase backtracking is not required in order for editing to occur. Thus, the 3′ end of the nascent RNA is expected to be in very close proximity to the mtRNAP active site. We hypothesize that the *Physarum* mtRNAP can adopt multiple active conformations, one of which is responsible for non-templated nucleotide addition.

8   Characterization of the *Physarum* Mitochondrial RNA Polymerase

Given that the *Physarum* mtRNAP is involved in editing, either directly (as the enzyme responsible for non-templated nucleotide addition) or indirectly (in editing site recognition and resumption of transcription from an added nucleotide), it was of interest to identify the gene that encodes it to further characterize this enzyme.

8.1   Cloning of the Mitochondrial RNA Polymerase

A two-step process was used to clone the nuclearly encoded mtRNAP from *Physarum polycephalum*. Our original clone was generated by PCR using DNA extracted from a cDNA library generously provided by Dr. Jennifer Dee as template (Bailey et al. 1992), and degenerate primers that were used by Cermakian et al. (1996) to survey the genes encoding mitochondrial RNA polymerases from *Triticum aestivum*, *Pycnococcus provasolii*, *Acanthamoeba castellanii*, *Isochrysis*, *Thraustochytrium aureum*, *Cryptomonas*, *Tetrahymena*, and *Naegleria* (kindly provided by Dr. Michael Gray). Sequence analysis of this ~500 bp cloned DNA fragment indicted that this region of the *Physarum* polymerase was closely related to other mitochondrial RNA polymerases, as well as being highly homologous to the single subunit RNA polymerases from bacteriophages T7, T3, and SP6 (Davanloo et al. 1984; Masters et al. 1987; Cermakian et al. 1996; Chen et al. 1996; Hedtke et al. 1997). Using the initial clone as a hybridization probe, we were then able to identify three larger clones from a *Physarum* cDNA library that we generated from polyA+ RNA. One of these longer clones was full length, based on 5′ RACE experiments (A.C Rhee, unpublished data).

8.2   Relationship of the Physarum Mitochondrial RNA Polymerase to Other Polymerases

The sequence of the *Physarum* mtRNAP is shown aligned with that of the bacteriophage T7 RNA polymerase (Genbank ascension number NP041960, Davanloo et al. 1984) in Fig. 5. The alignment shown is based on a multiple alignment using the mitochondrial RNA polymerases from *Saccharomyces cerevisiae* (Masters et al. 1987), *Arabidopsis thaliana* (Hedtke et al. 1997), and *Neurospora crassa* (Chen et al. 1996) in addition to the bacteriophage T7 enzyme (the sequences of the longer mitochondrial enzymes are not shown to minimize cluttering the figure). The polymerase domains of each of these enzymes are quite similar, with high conservation of the previously identified A, B, and C motifs (Fig. 6; Joyce and Steitz 1995), including the catalytic aspartic acid residues (underlined in Figs. 5 and 6) and the invariant lysine in motif B (arrow in Fig. 6). In contrast, the N-terminal regions of mtRNAPs from different species have diverged significantly; this region of the *Physarum* mtRNAP shows little homology to either other mitochondrial enzymes or bacteriophage polymerases (Fig. 5; other data not shown).

Multiple high-resolution crystal structures of the bacteriophage T7 RNA polymerase have been solved, including structures in the absence and presence of DNA, nucleotides, and pyrophosphate. The structures representing the elongating

```
Pp  MPLTHLPVFLRHSGSVHAHYLLNHHHVSAIPSSSLLCSSRSAPFFGFGVRVISIRAVQVQ

Pp  EPGGQFPAVDPKRVPPPNFKSQFDFYRDKNARDIAQELKEDADEKDEDNMTDEDRYDIYE
T7  ---------------------------------MNTINIAKNDFSD-------IELAA

Pp  DDTSYPSNDKVQPHHRDERVVLREAEVSIEAAEDFAKREKKEQWWRMTQEELAKYQLQAD
T7  IPFNTLADHYGERLAR-EQLALEH--------ESYEMGEAR------FRKMFERQLKAG

Pp  LELNAIDQATEYYRKALDGILRARQGASLGPTRRIIDSWYGPFRDSINAAVRKMSARASS
T7  EVAD-----NAAAKPLITTLLPKMIARIND--------WFEEVKAKRG----------K

Pp  HSLVAN-LKLLSSDKLPVITLHQTLGMLLANTEGVPFMQVAFAVGRAVQAEVNFERMKA-
T7  RPTAFQFLQEIKPEAVAYITIKTTLACLTSA-DNTTVQAVASAIGRAIEDEARFGRIRDL

Pp  ---------EDKQAFRSLLKSKSGITVTLVNIKAKYTL----KKHDWDSKTIVQLGAFLL
T7  EAKHFKKNVEEQLNKRVGHVYKKAFMQVVEADMLSKGLLGGEAWSSWHKEDSIHVGVRCI

Pp  KHLMLSAKIEKNVDKVLSKANNLAVSSSERQLADDEYVSAFVHYYKFYRQGSTTRRQGLL
T7  EMLIESTGMVSLHRQNAG----VVGQDSE-------------------------TI

Pp  KCHENIFKYVDDGHVIKEVLDARLLPMVVKPVPWVSPTVGGYLYV---PSVLMRAQEHQA
T7  ELAPEYAEAIATRAGALAGISPMFQPCVVPPKPWTGITGGGYWANGRRPLALVRTH-SKK

Pp  QYASLWNADLTTVFAALNALGETGWRVNEEVYHIVKKV--WDDGGGLAALPTRTDITFPD
T7  ALMRYEDVYMPEVYKAINIAQNTAWKINKKVLAVANVITKWKHCP-VEDIPAIEREELPM

Pp  PPDDLDTNDESKRTWKKTERKIRQLNNDLNSLRCDTTYKLSVAETLLHMD-FYLPHNVDF
T7  KPEDIDMNPEALTAWKRAAAAVYRKDKARKSRRISLEFMLEQANKFANHKAIWFPYNMDW

Pp  RGRAYPIPPHLNQLGSDMCRGLLVFKEKKPLGPTGLRWLKIHLANLYGVDKVPFDARVNF
T7  RGRVYAVS-MFNPQGNDMTKGLLTLAKGKPIGKEGYYWLKIHGANCAGVDKVPFPERIKF

Pp  VDKNIDKVFDSADHPLDGEQWWVGADDPWQCLGTCFELAKALRSPNPEEFLSCMPVHQDG
T7  IEENHENIMACAKSPLEN-TWWAEQDSPFCFLAFCFEYAGVQHHG--LSYNCSLPLAFDG
                                                         MOTIF A
Pp  SCNGLQHYAALGGDEIGGRKVNLVPSDAPQDVYSGVATVVAKRVHEDAIAGHE-------
T7  SCSGIQHFSAMLRDEVGGRAVNLLPSETVQDIYGIVAKKVNEILQADAINGTDNEVVTVT

Pp  -----------------LGKLLDGKLDRKVVKQTVMTSVYGVTYIGARLQIHNALSDKD
T7  DENTGEISEKVKLGTKALAGQWLAYGVTRSVTKRSVMTLAYGSKEFGFRQQVLEDTIQPA
                         MOTIF B
Pp  IE---WRDDDQLYHASAYITNHTFEALNQMFLGARNIMKWLADCARIIA---KNKDP---
T7  IDSGKGLMFTQPNQAAGYMAKLIWESVSVTVVAAVEAMNWLKSAAKLLAAEVKDKKTGEI

Pp  ------VAWVTPLGLPIVQPYVKNQQFLVKTVVQKVLLSDP----RKDLPVNTIKQRSAF
T7  LRKRCAVHWVTPDGFPVWQEYKKPIQTRLNLMFLGQFRLQPTINTNKDSEIDAHKQESGI

Pp  PPNYVHSLDSCHMMLTAIECQKK-GI-SYTSVHDSYWTHACTVDQMNVILRDQFVELH-S
T7  APNFVHSQDGSHLRKTVVWAHEKYGIESFALIHDSFGTIPADAANLFKAVRETMVDTYES
                                       MOTIF C
Pp  QPLLHRLREWFIKRYGKDGRVEFPPVPERGPLDLQVVKQSKYFFH
T7  CDVLADFYDQFADQLHESQLDKMPALPAKGNLNLRDILESDFAFA
```

Fig. 5 Alignment of the predicted translation product of the *Physarum* mitochondrial RNA polymerase gene with the bacteriophage T7 RNA polymerase. Note that alignments are based on Clustal alignments with the mitochondrial RNA polymerases from *Saccharomyces cerevisiae* (NP116617), *Arabidopsis thaliana* (CAA70583), and *Neurospora crassa* (P38671) in addition to the bacteriophage T7 enzyme

SOURCE	MOTIF A	MOTIF B	MOTIF C
	★	⇓	★
Pp	PVHQDGSCNGLQHY	RKVVKQTVMTSVY	VHDSYWT
Dd	PIHQDGTCNGLQHY	RKLVKQTVMTSVY	VHDSYWT
Af	PIHQDGSCNGLQHY	RKIVKQTVMTNVY	VHDSFWT
Nc	PIHQDGTCNGLQHY	RKVVKQTVMTNVY	VHDSFWT
Nt	PVHQDGSCNGLQHY	RKLVKQTVMTSVY	VHDSYWT
Sc	PVHQDGTCNGLQHY	RKVVKQTVMTNVY	VHDSYWT
Sp	PIQQDGTCNGLQHY	RSVVKPTVMTNVY	VHDSYWT
Mm	PVHQDGSCNGLQHY	RKVVKQTVMTVVY	VHDCFWT
Hs	PVHQDGSCNGLQHY	RKVVKQTVMTVVY	VHDCYWT
Tb	PVAVDGSYNGLQHY	RKTIKRPIMTQVY	VHDSYWT
Nt cp	PIHQDGSCNGLQHY	RKLVKQTVMTSVY	VHDSFWT
At cp	PIHQDGSCNGLQHY	RKLVKQTVMTSVY	VHDSYWT

Fig. 6 Alignment of conserved polymerase motifs A, B, and C from the single subunit bacteriophage polymerases T7 and T3 with mitochondrial and chloroplast RNA polymerases. Motifs A, B, and C are taken from Joyce and Steitz (1995). Abbreviations and Genbank accession numbers are as follows: T7 (bacteriophage T7), NP041960; T3 (bacteriophage T3), NP523301; Pp (*Physarum polycephalum*), accession number pending; Dd (*Dictyostelium discoideum*), XP647347; Af (*Aspergillus fumigatus*), EAL93858; Nc (*Neurospora crassa*), P38671; At (*Arabidopsis thaliana*), CAA70583 (mt, mitochondrial) and CAA69972 (cp, chloroplast); Nt (*Nicotiana tabacum*), CAC95022 (mt, mitochondrial) and CAC95027 (cp, chloroplast); Sc (*Saccharomyces cerevisiae*), NP116617; Sp (*Schizosaccharomyces pombe*), CAB16197; Mm (*Mus musculus*), NP766139; Hs (*Homo sapiens*), AAB58255; Tb (*Trypanosoma brucei*), XP828653

form of T7 polymerase (Yin and Steitz 2002, 2004; Tahirov et al. 2002; Temiakov et al. 2004) are of particular interest in terms of the *Physarum* mtRNAP, given that editing occurs during transcriptional elongation in *Physarum* mitochondria. Based on these crystal structures, the DNA sequences required for editing are likely to be either in direct contact with the *Physarum* mtRNAP or in very close proximity to the enzyme. In addition, the RNA–DNA hybrid, which includes the 3′ end of the RNA that is the substrate for editing in *Physarum* mitochondria, is almost completely buried within the T7 RNA polymerase transcription elongation complex (Tahirov et al. 2002). Given the strong conservation of the active site residues (Fig. 6), we predict that the RNA–DNA hybrid in *Physarum* transcription elongation complexes is similarly buried, consistent with our hypothesis that the *Physarum* mtRNAP is the enzyme responsible for non-templated nucleotide addition.

9 Future Directions

Although much has been learned regarding the mechanism of insertional editing in *Physarum* mitochondria, significant questions remain. How are the sites of nucleotide insertion (or deletion) determined? What specifies which nucleotide or nucleotides to add (or remove) at a given site? What factors are required? At what point in the editing cycle do they act? How are the transitions between templated transcription and editing (and visa versa) accomplished? How are the extra nucleotides accommodated in the RNA–DNA hybrid during templated transcription? It is hoped that current biochemical and structural approaches will provide insights into this intriguing form of gene expression, as well as lead to a better understanding of transcription elongation mechanisms in general.

Acknowledgements The authors wish to thank Greg Pietz, who cloned the initial portion of the *Physarum* mitochondrial polymerase, Drs. Michael Gray and Jennifer Dee, who provided the primers and cDNA library, respectively, used to generate this clone, and present and former lab members who contributed, directly or indirectly, to this work. This work was supported by NIH grant GM54663 (to JMG) and NIH training grant #T32 AG00105 (to ACR).

References

Antes T, Costandy H, Mahendran R, Spottswood M, Miller D (1998) Insertional editing of mitochondrial tRNAs of *Physarum polycephalum* and *Didymium nigripes*. Mol Cell Biol 18:7521–7527

Bailey J, Solnica-Krezel L, Lohman K, Dee J, Anderson RW, Dove WF (1992) Cellular and molecular analysis of plasmodium development in *Physarum*. Cell Biol Int Rep 16:1083–1090

Bundschuh R (2004) Computational prediction of RNA editing sites. Bioinformatics 20:3214–3220

Byrne EM, Gott JM (2002) Cotranscriptional editing of *Physarum* mitochondrial RNA requires local features of the native template. RNA 8:1174–1185

Byrne EM, Gott JM (2004) Unexpectedly complex editing patterns at dinucleotide insertion sites in *Physarum* mitochondria. Mol Cell Biol 24:7821–7828

Byrne EM, Stout A, Gott JM (2002) Editing site recognition and nucleotide insertion are separable processes in *Physarum* mitochondria. EMBO J 21:6154–6161

Cermakian N, Ikeda TM, Cedergren R, Gray MW (1996) Sequences homologous to yeast mitochondrial and bacteriophage T3 and T7 RNA polymerases are widespread throughout the eukaryotic lineage. Nucleic Acids Res 24:648–654

Chen B, Kubelik AR, Mohr S, Breitenberger CA (1996) Cloning and characterization of the *Neurospora crassa cyt*-5 gene – A nuclear-coded mitochondrial RNA polymerase with a polyglutamine repeat. J Biol Chem 271:6537–6544

Cheng YW, Gott JM (2000) Transcription and RNA editing in a soluble *in vitro* system from *Physarum* mitochondria. Nucleic Acids Res 28:3695–3701

Cheng YW, Visomirski-Robic LM, Gott JM (2001) Non-templated addition of nucleotides to the 3′ end of nascent RNA during RNA editing in *Physarum*. EMBO J 20:1405–1414

Davanloo P, Rosenberg AH, Dunn JJ, Studier FW (1984) Cloning and expression of the gene for bacteriophage T7 RNA polymerase. Proc Natl Acad Sci USA 81:2035–2039

Gott JM, Visomirski-Robic LM (1998) RNA editing in *Physarum* mitochondria. In: Grosjean H, Benne R (eds) Modification and editing of RNA. ASM Press, Washington, DC, pp 395–411

Gott JM, Visomirski LM, Hunter JL (1993) Substitutional and insertional RNA editing of the cytochrome c oxidase subunit 1 mRNA of *Physarum polycephalum*. J Biol Chem 268:25483–25486

Gott JM, Parimi N, Bundschuh R (2005) Discovery of new genes and deletion editing in *Physarum* mitochondria enabled by a novel algorithm for finding edited mRNAs. Nucleic Acids Res 33:5063–5072

Hedtke B, Borner T, Weihe A (1997) Mitochondrial and chloroplast phage-type RNA polymerases in *Arabidopsis*. Science 277:809–811

Horton TL, Landweber LF (2000) Evolution of four types of RNA editing in myxomycetes. RNA 6:1339–1346

Johnson DF, Poksay KS, Innerarity TL (1993) The mechanism for apo-B mRNA editing is deamination. Biochem Biophys Res Commun 195:1204–1210

Joyce CM, Steitz TA (1995) Polymerase structures and function: variations on a theme? J Bacteriol 177:6321–6329

Laforest MJ, Roewer I, Lang BF (1997) Mitochondrial tRNAs in the lower fungus *Spizellomyces punctatus*: tRNA editing and UAG 'stop' codons recognized as leucine. Nucleic Acids Res 25:626–632

Lonergan KM, Gray MW (1993) Editing of transfer RNAs in *Acanthamoeba castellanii* mitochondria. Science 259:812–816

Mahendran R, Spottswood MR, Miller DL (1991) RNA editing by cytidine insertion in mitochondria of *Physarum polycephalum*. Nature 349:434–438

Mahendran R, Spottswood MS, Ghate A, Ling ML, Jeng K, Miller DL (1994) Editing of the mitochondrial small subunit rRNA in *Physarum polycephalum*. EMBO J 13:232–240

Masters BS, Stohl LL, Clayton DA (1987) Yeast mitochondrial RNA polymerase is homologous to those encoded by bacteriophages T3 and T7. Cell 51:89–99

Miller D, Mahendran R, Spottswood M, Costandy H, Wang S, Ling ML, Yang N (1993) Insertional editing in mitochondria of *Physarum*. Semin Cell Biol 4:261–266

Miller ML, Antes TJ, Qian F, Miller DL (2006) Identification of a putative mitochondrial RNA polymerase from *Physarum polycephalum*: characterization, expression, purification, and transcription in vitro. Curr Genet 49:259–271

Nair DT, Johnson RE, Prakash S, Prakash L, Aggarwal AK (2004) Replication by human DNA polymerase-iota occurs by Hoogsteen base-pairing. Nature 430:377–380

Rajasekhar VK, Mulligan RM (1993) RNA editing in plant mitochondria: [alpha]-phosphate is retained during C-to-U conversion in mRNAs. Plant Cell 5:1843–1852

Rundquist BA, Gott JM (1995) RNA editing of the *col* mRNA throughout the life cycle of *Physarum polycephalum*. Mol Gen Genet 247:306–311

Tahirov TH, Temiakov D, Anikin M, Patlan V, McAllister WT, Vassylyev DG, Yokoyama S (2002) Structure of a T7 RNA polymerase elongation complex at 2.9 Å resolution. Nature 420:43–50

Takano H, Abe T, Sakurai R, Moriyama Y, Miyazawa Y, Nozaki H, Kawano S, Sasaki N, Kuroiwa T (2001) The complete DNA sequence of the mitochondrial genome of *Physarum polycephalum*. Mol Gen Genet 264:539–545

Temiakov D, Patlan V, Anikin M, McAllister WT, Yokoyama S, Vassylyev DG (2004) Structural basis for substrate selection by T7 RNA polymerase. Cell 116:381–391

Tomita K, Fukai S, Ishitani R, Ueda T, Takeuchi N, Vassylyev DG, Nureki O (2004) Structural basis for template-independent RNA polymerization. Nature 430:700–704

Vidal S, Curran J, Kolakofsky D (1990) A stuttering model for paramyxovirus P mRNA editing. EMBO J 9:2017–2022

Visomirski-Robic LM, Gott JM (1995) Accurate and efficient insertional RNA editing in isolated *Physarum* mitochondria. RNA 1:681–691

Visomirski-Robic LM, Gott JM (1997a) Insertional editing in isolated *Physarum* mitochondria is linked to RNA synthesis. RNA 3:821–837

Visomirski-Robic LM, Gott JM (1997b) Insertional editing of nascent mitochondrial RNAs in *Physarum*. Proc Natl Acad Sci USA 94:4324–4329

Wang SS, Mahendran R, Miller DL (1999) Editing of cytochrome b mRNA in *Physarum* mitochondria. J Biol Chem 274:2725–2731

Xiong Y, Steitz TA (2004) Mechanism of transfer RNA maturation by CCA-adding enzyme without using an oligonucleotide template. Nature 430:640–645

Yin YW, Steitz TA (2002) Structural basis for the transition from initiation to elongation transcription in T7 RNA polymerase. Science 298:1387–1395

Yin YW, Steitz TA (2004) The structural mechanism of translocation and helicase activity in T7
    RNA polymerase. Cell 116:393–404
Yoshizawa S, Fourmy D, Puglisi JD (1999) Recognition of the codon-anticodon helix by ribos-
    omal RNA. Science 285:1722–1725
Yue D, Weiner AM, Maizels N (1998) The CCA-adding enzyme has a single active site. J Biol
    Chem 273:29693–29700

RNA Editing in Plant Mitochondria

Mizuki Takenaka (✉), Johannes A. van der Merwe, Daniil Verbitskiy, Julia Neuwirt, Anja Zehrmann, and Axel Brennicke

> *"All our science, measured against reality, is primitive and*
> *childlike – and yet it is the most precious thing we have."*
> Albert Einstein

Abstract RNA editing in plant mitochondria alters more than 400 cytidines to uridines in flowering plants. In other plants such as ferns and mosses, the reverse reaction is observed at almost equal frequency. In the last few years, the development of transfection systems with isolated mitochondria and of in vitro systems with mitochondrial extracts has considerably improved our understanding of the parameters of site recognition. However, the biochemistry and the enzymes involved are still open questions. We here summarize our present knowledge of RNA editing as an essential part of RNA maturation in flowering plant mitochondria.

Molekulare Botanik, Universität Ulm, 89069 Ulm, Germany; mizuki.takenaka@uni-ulm.de

H.U. Göringer (ed.), *RNA Editing. Nucleic Acids and Molecular Biology 20*
© Springer-Verlag Berlin Heidelberg 2008

1   Introduction

In mitochondria and plastids of almost all the land plants examined, the sequences of many transcripts are altered by posttranscriptional conversion of C to U and U to C (Covello and Gray 1989; Gualberto et al. 1989; Hiesel et al. 1989). In plastids, including chloroplasts of flowering plants, in total only about 35 Cs are found to be deaminated to Us, details of which are described in the chapter by Masahiro Sugiura (this volume).

In mitochondria, by extrapolation from the RNAs analysed to date, about 400–500 C to U changes are expected. In *Arabidopsis thaliana*, *Brassica napus* and *Oryza sativa*, the entire set of editing sites in protein-coding regions has been investigated, amounting to 441, 427 and 491 editing sites per mitochondrial transcriptome in these species respectively (Giegé and Brennicke 1999; Notsu et al. 2002; Handa 2003). The total number of C to U changes will be somewhat higher, since several events have been documented in non-protein-coding regions such as tRNAs and introns, as well as in leader and trailer regions. However, these are comparatively few editing sites; by extrapolation, about 20–50 can be expected in a given transcriptome.

RNA editing is required for gene expression in many of the systems discussed in this volume. Often, the genomic information encoding an open reading frame or a tRNA is cryptic or incomplete, and will not yield a functional product. Thus, the corresponding genetic system is dependent on RNA editing for its biological optimization and eventually for survival. In the plant mitochondrial systems, this process is likewise essential for the synthesis of functional proteins which, after editing, exhibit closer sequence conservation with their homologs in other systems. In addition, RNA editing in plant mitochondria ensures that several affected tRNAs can fold correctly and become functional (Marchfelder and Binder 2004). Thus, RNA editing in plant mitochondria is an essential step of RNA maturation without which no working respiratory chain and no functional mitochondria can be produced and maintained in the cell.

In this overview, we will first summarize the observations of RNA editing in plant mitochondria in the steady-state RNA population, and the consequences of editing. In the second part, we will look at recent approaches trying to characterize the biochemistry involved and to identify the parameters of site recognition.

2   C to U and U to C Changes and Their Distribution in the Plant Kingdom

RNA editing sites in flowering plant mitochondria involve almost exclusively C to U changes, in which the C in the pre-mRNA is deaminated to a U moiety (Fig. 1). The more than 400 RNA editing sites are found largely in the coding regions of mRNAs and occur less frequently in introns and other non-translated regions. In some cases, RNA editing in tRNA molecules restores essential base-pairings. In

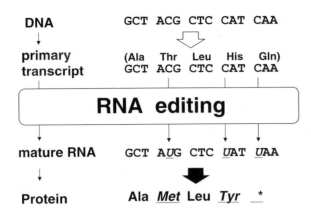

Fig. 1 RNA editing in mitochondria of flowering plants alters cytosines (C) in the primary transcript to uridines (U) in the mature mRNA. The edited mRNA specifies an amino acid sequence different from the protein predicted from the genomic DNA and the primary transcript. Several examples of such amino acid changes are given. In plant mitochondria of a given angiosperm species, about 400 to 500 such events are observed. These have to be specified and differentiated from cytosines which are not altered by RNA editing

these instances, only complete RNA editing allows correct folding and further maturation and processing of the tRNA precursors. For as yet unclear reasons, the ribosomal RNAs in plant organelles appear to undergo little RNA editing, if any.

In flowering plant species, U to C conversions are very rare events. In *Arabidopsis thaliana*, *Brassica napus* and *Oryza sativa*, none have been identified (Giegé and Brennicke 1999; Notsu et al. 2002; Handa 2003), while only one each have been reported in the mitochondrial RNAs of the spermatophyte species wheat and *Oenothera* (Gualberto et al. 1990; Schuster et al. 1990).

In non-flowering plants, the reverse reaction – the amination of U to C – is much more common and can reach the level of the C to U editing events, for example, in hornworts (Malek et al. 1996; Knoop 2004). The only land plants in which no editing has been observed at all is the liverwort *Marchantia polymorpha* (Oda et al. 1992) and closely related species of the Marchantiales (Malek et al. 1996). Other liverworts, for example, of the order Metzgeriales, are edited at least in their mitochondria (Malek et al. 1996). Green algae, including the closest relatives to the land plants of the order Charales, do not show any RNA editing (Turmel et al. 2003). One possible date for the establishment of RNA editing in plant mitochondria is thus in the early evolution of land plants, since the Marchantiales are now considered to branch near the very root of the land plant tree (Malek et al. 1996; Knoop 2004).

Intriguing, and with far-reaching consequences for possible biochemical reactions, is the observation of frequent U to C aminations among the mosses and ferns, for example, in hornwort species and the Lycopodiopsida (Knoop 2004; Groth-Malonek and Knoop 2005). From previously characterized cytidine deaminases and cytidine synthases, it is difficult to envisage a single enzyme performing both reactions (discussed further below). The asymmetric distribution of editing sites between coding and non-coding regions in RNAs suggests that there has to be some selective pressure on maintaining an editing site, which would otherwise be lost. The necessity to produce functional proteins with certain amino acid sequences may contribute to this selective pressure in open reading frames, where a U at a given editing position may be required for a specific codon identity. In introns, RNA editing is in some cases a prerequisite for proper folding of essential secondary structures and, thus, efficient intron excision (Carillo and Bonen 1997), suggesting that the requirement of a functional protein product exercises evolutionary pressure on this editing event to occur.

The present-day requirement for RNA editing has most likely evolved after the establishment of the biochemical potential to posttranscriptionally modify certain nucleotide positions. This scenario was first clearly outlined in the review by Covello and Gray (1993), in which the authors suggest that, as a first step, a deaminating enzyme mutates to also accept polynucleotide templates and that thereafter the DNA is free to mutate in certain positions. Once one such mutation had occurred which required a compensating correction in the RNA, RNA editing became established and needed to be functionally maintained. This scenario was aptly named the bureaucracy model (Mulligan 2004), in analogy to the various processes in administrations which introduce often unnecessary complexities with the sole aim of making themselves indispensable.

3   Consequences and Advantages of RNA Editing

RNA editing per se is rather extravagant and costly because, without exception, all of the RNA editing events described would be rendered unnecessary if the sequence of the mature mRNA were encoded in the genome, as is the case for bacteria. The question thus arises whether there is a hidden biological significance of RNA editing. Such a camouflaged agenda might include a regulatory control of gene expression at the posttranscriptional level where, for example, the introduction of an AUG translational start codon could rapidly provide an mRNA for translation without requiring the complete de novo synthesis of the transcript.

Another advantage of RNA editing could be the potential of synthesizing two or more distinct products from a single gene, as exemplified by the viral editing systems or in the tissue-specific apoB editing, where two lipid carrier proteins with different properties are derived from one and the same genomic coding region (Greeve et al. 1991; Chang et al. 1998; see chapter by Speijer, this volume). Editing in the GluR receptor channels results in an increase in the family of variant proteins available for the combinatory assembly of the channels, and allows fine-tuning of the chemo-electric potential transmission properties (see chapter by Jantsch and Öhman, this volume).

Partially edited mRNAs are found in polysomal fractions of plant mitochondria and are apparently translated into a family of variant proteins (Mulligan 2004). However, only one type of protein sequence appears to be incorporated into the polypeptide complexes of the respiratory chain. These integrated polypeptide sequences generally correspond to the protein specified by the fully edited mRNA (excluding the sporadic editing events), and not to any of the intermediates (Phreaner et al. 1996; Mulligan 2004). The polypeptide sequences predicted from the fully edited mRNA are better conserved with the homologs in other organisms than the proteins encoded by the genomic sequences or by any of the editing intermediates and, thus, may be selected for their conserved physiological and biochemical functionality. Indeed, it is unlikely that polypeptides synthesized from unedited RNAs would function properly; such proteins would rather impair the efficiency of mitochondrial respiration. This has been investigated by importing an ATP9 protein synthesized from an (unedited) genomic mitochondrial sequence inserted into the nuclear genome of tobacco. This protein indeed disturbs mitochondrial function, presumably by competition with the endogenous native, mitochondrially encoded protein. The resulting dysfunction manifests itself in the form of problematical pollen development, leading to male sterility (Hernould et al. 1993; Zabaleta et al. 1996).

Even if the coding capacity of a gene could be increased by using partially edited mRNAs as templates to produce protein variants without increasing the number of genes in plant mitochondria, this presumed advantage may be illusionary because additional polypeptides are required for the editing process. RNA editing thus appears to be a priori a biologically selfish process, which has become established (by chance) and now has to be perpetuated.

On the other hand, the mitochondrial genome may have 'welcomed' the establishment of RNA editing as a means of protecting its genome and genes against transfer to the nucleus. Since there is no evidence of editing in the nucleus/cytoplasm of plants, a mitochondrial gene requiring editing could not immediately become functional upon transfer to the nuclear genome. A successful transfer would require the respective coding region to be reverse transcribed from a fully edited mitochondrial mRNA molecule. The cDNA sequence could code for the correct protein sequence once inserted and activated in the nuclear genome.

Plant mitochondria (and plastids) employ the normal genetic code, while mitochondria in many other species use variants of the codon translation system. The coincidence of identical decoding systems with the same translational cipher in mitochondria and in the cytoplasmic system is probably unrelated to the advent of RNA editing and may rather be connected with the import of various tRNA species from the cytoplasm into mitochondria. Since several of these tRNAs are shared between the mitochondrion and cytoplasm, a stringent selection pressure is imposed which maintains the standard code in both compartments.

4   Biochemistry of RNA Editing in Plant Mitochondria

Since no nucleotide insertions or other changes of nucleotide identities, as found in kinetoplasts (Benne et al. 1986; Chang et al. 1998; Hajduk and Sabatini 1998; see chapters by Carnes and Stuart, Göringer et al., and Ochsenreiter and Hajduk, all this volume) or in the slime mould *Physarum polycephalum* (Mahendran et al. 1991; Gott and Visomirski-Robic 1998; see chapter by Gott and Rhee, this volume), have been observed in plants, the most parsimonious explanation of the enzymatic mechanism of RNA editing would be to invoke the activity of an RNA-specific C-deaminase analogous to the APOBEC-1 enzyme described in the apoB editing of Mammalia (Greeve et al. 1991; Navaratnam et al. 1998). Evidence gathered to date does indeed point to a deamination reaction being involved in the C to U conversions in plant mitochondria (Rajasekhar and Mulligan 1993; Yu and Schuster 1995; Mulligan 2004).

Experiments with in vitro and in organello assays investigating the fate of the sugar-phosphate backbone as well as the nucleotide base have shown that both are retained in the editing process (Rajasekhar and Mulligan 1993; Yu and Schuster 1995). These observations exclude nucleotide excision and base exchange as possible reactions, implying deamination or transamination as resulting in the C to U change seen in flowering plant mitochondria.

Transamination reactions are common in the various pathways of amino acid biosynthesis, where they require a limited number of compounds as amino group acceptors. Several of these have been tested for their effects on in vitro editing reactions (e.g. oxaloacetate and alpha-ketoglutarate) but none showed any influence on the efficiency of the reaction (Mizuki Takenaka, unpublished data). Furthermore, the in vitro systems developed for plant mitochondria (as well as for chloroplasts; see chapter by Sugiura, this volume) are extensively dialysed prior to use, which should quantitatively remove all small molecules below 10 kDa such as oxaloacetate and alpha-ketoglutarate. If these or similar compounds were required as amino group receptors, then the RNA editing reaction should not be possible in vitro without their addition to the assay. Such observations make a related transamination reaction unlikely to be involved in the C to U deamination observed in flowering plant mitochondria.

This leaves us with a modified cytidine deaminase as the most likely candidate for the responsible enzyme. In the fully sequenced nuclear genome of *Arabidopsis thaliana*, a family of nine genes has been described, at least one of which is targeted to mitochondria (Faivre-Nitschke et al. 1999). This enzyme could be involved in the RNA editing reaction, while others would perform the classic role of cytidine deaminase in the nucleotide biosynthesis pathway from cytidine towards uridine.

However, there are two arguments against the participation of such straightforward cytidine deaminases. The first is based on experimental evidence which suggests that no accessible zinc ions are required for the reaction. While all classic cytidine deaminases contain reaction-centre bound zinc and can be incapacitated by specific zinc chelators such as 1,10 o-phenanthroline, the in vitro RNA editing reaction is not inhibited in plant mitochondrial extracts by the addition of this compound (Takenaka and Brennicke 2003). The bacterial cytidine deaminases as well as the apo-B cytidine deaminase are both completely blocked by addition of this chelator (Carter 1998). Similarly, in chloroplast in vitro assays, the zinc chelator has some detrimental effect on the editing reaction (Hegemann et al. 2005). The unaffected plant mitochondrial enzyme must have evolved along a different track to protect or to not require the central zinc ion. Alternatively, a completely different deaminating enzyme with an entirely distinct evolutionary origin is involved.

The second argument against the involvement of such cytidine deaminases derives from the 'reverse' editing events of U to C frequently observed in ferns and mosses. Based on energetic considerations, cytidine deaminases are unlikely to be able to catalyse the reverse reaction of adding an amino group to a U residue to generate a C. Of course, a second unrelated enzyme could be responsible for these reverse reactions; the obvious solution would, however, be a single enzyme system able to perform both reactions. In the hornwort and related plant species, this would be a variant protein which is more prone to catalyze the U to C reaction, while the distinctly evolved enzyme in flowering plants is less suited to facilitate this reaction.

Another possible evolutionary source of an enzyme able to perform amination reactions might be the family of CTP synthases. However, these do not support cytidine deamination. None of the class of CTP synthase enzymes characterised for mononucleotide amination from UTP to CTP is able to catalyze the deamination step. However, a highly evolved enzyme might have become adapted to enhance this reaction more efficiently. Another problem with the CTP synthases arises from their requirement for cofactors as amino group donors, usually ammonia or glutamine; as detailed above for the transaminases, these should have been quantitatively eliminated from the editing-competent in vitro lysates by dialysis. An argument in favour of an origin of the editing enzyme from within the CTP synthase family is that, in plant mitochondrial in vitro assays, a high concentration of ATP is required and, in chloroplast lysates, additional ATP greatly enhances the performance. In the assayed C to U changes, of course, no amino group donor would be required – rather, an acceptor might be necessary.

The enzymatic reaction will be rapidly clarified once the involved enzyme has been identified through biochemical or genetic analysis. This seems to be the imminent task of plant mitochondrial RNA editing research in the near future, which hopefully will be accomplished as soon as experimental protein identifications have progressed in that direction.

5 Involvement of an RNA Helicase

The abovementioned requirement for ATP in the in vitro reactions may be due to auxiliary reactions, rather than a direct need of energy for the actual deamination of C to U. Intriguing in this respect is the observation that, in plant mitochondrial in vitro assays from pea as well as from cauliflower, ATP can be substituted by any of the triphosphates, including any of the dNTPs. A separate but necessary auxiliary function with such a broad spectrum of nucleotidetriphosphate acceptance may be an RNA helicase activity, since several members of this class of enzymes have been shown to tolerate such a broad spectrum of nucleotides. RNA helicases are involved in many different processes of RNA maturation and could possibly also be required for efficient RNA editing in plant mitochondria.

An RNA helicase could resolve secondary structures in the template mRNAs and present the RNA molecule in a linear form accessible to the specific *trans*-acting factors, which are most likely proteins. Alternatively, the helicase could clear the RNA template from unspecific binding proteins which adhere to any RNA, and protect it from nucleases and other deleterious influences (Fairman et al. 2004). In this function, the helicase would move along the RNA molecule and dissociate the bound unspecific proteins, a process requiring the energy provided by the nucleotidetriphosphate. The cleaning of the RNA would free the recognition site and the nucleotide to be edited and, thus, allow access to the editing activity.

6 Investigations of the Specificity Determinants

RNA editing in plant organelles is a posttranscriptional process. This was suggested by the identification of potential RNA editing intermediates and, of course, by the successful establishment of in vitro systems which edit added RNA molecules. In partially edited transcripts, some Cs have been changed into Us while, at other potential editing sites, the Cs encoded by the genome are (still) present. These partially edited RNA molecules show no order of editing along the RNA molecule, suggesting that the editosome complex does not scan the RNA molecule along its linear length but rather selects certain sites or regions in a hit-and-run mode, and releases the transcript again after editing.

In vitro experiments addressed to determine whether partially edited transcripts are true intermediates, and can indeed serve as substrates for further rounds of RNA editing, employed substrates in which either of two closely spaced editing sites was already modified (Verbitskiy et al. 2006). Both substrates are accepted with efficiencies comparable to completely unedited RNA molecules, showing that partially edited RNAs can indeed serve as substrates for further rounds of RNA editing. This result further supports the hit-and-run model, in which specific regions of the RNA molecule are addressed separately.

It is obvious that each editing site needs to be identified individually to differentiate a C to be edited from a C-moiety which should not be altered. The potentially discriminating information can reside only in the nucleotide sequence context surrounding a given editing site. Theoretical considerations suggest that the minimal requirement to unambiguously identify a nucleotide sequence motif in the total mitochondrial genome complexity could be as little as 10 nucleotides. In silico analyses of large numbers of editing sites, e.g. all the 441 sites in the protein-coding regions of the *Arabidopsis thaliana* mitochondrial RNA, did not reveal unique motif(s) of comparable size common to all or even many of the sites (Giegé and Brennicke 1999). The comparative analyses of site vicinities only confirmed preferences for certain nucleotide identities immediately adjacent to the edited C. For example, at the position immediately upstream of the edited C, guanosines are rarely observed, suggesting that the editing enzyme (complex) sterically prefers pyrimidines, and might be hindered particularly by a G, at this position.

Genome internal recombinations have created in vivo transcripts in which a given editing site of a bona fide open reading frame is duplicated with few surrounding nucleotides of the original context (Lippok et al. 1994; Kubo and Kadowaki 1997). Such natural chimeric constructs suggest that most of the specificity determinants are located upstream and that as little as three conserved nucleotides should be sufficient downstream to mark a given editing site (Lippok et al. 1994).

The requirements of size and location of the specificity-conferring nucleotide context relative to the edited nucleotide in plant mitochondria have recently been experimentally investigated in detail by in organello and in vitro approaches.

In the 15 years since RNA editing was first recognized in plant mitochondria and chloroplasts as a posttranscriptional process altering mostly C to U nucleotide identities in mRNAs and tRNAs, progress towards elucidating the enzymes and the specificity recognition has been impeded mostly by the lack of efficient in vitro systems. In vivo analysis of transgenic chloroplasts has brought important insights into the structure and extension of *cis*-elements, but this approach is difficult to extend towards a biochemical characterization and the identification of the corresponding *trans*-factors (Chaudhuri et al. 1995; Chaudhuri and Maliga 1996; Bock et al. 1996, 1997; Reed et al. 2001; Chateigner-Boutin and Hanson 2002).

With the development of reliable in vitro activities for chloroplasts (Hirose and Sugiura 2001; Miyamoto et al. 2002, 2003) and for pea and cauliflower mitochondria (Takenaka and Brennicke 2003), as well as the establishment of wheat and maize in organello editing systems (Farré and Araya 2001) in the past few years, characterization of the *cis*-requirements at individual sites has accelerated considerably (Table 1).

6.1   *In Organello RNA Editing*

An in organello assay has been successfully established by the group of Alejandro Araya for wheat mitochondria (Farré and Araya 2001). Its general feasibility has been confirmed by transfer to maize in the laboratory of Frank Kempken (Staudinger and Kempken 2003). Briefly, DNA molecules with appropriate promoter and RNA stability structures are introduced into intact isolated mitochondria by electroporation. The DNA is transcribed in the organelles, and the RNA is processed by correct intron excision and RNA editing at many of the in vivo observed sites.

Several of these active sites have recently been investigated by mutating various positions around the edited nucleotides. These investigations show that 16

Table 1 Overview of several different procedures developed to analyse RNA editing in plant mitochondria

Procedure	Advantages	Limitations	Reference
In organello	Closest to the native in vivo situation. Can see the 'whole' picture of many sites	Difficult to alter biochemical conditions. Allows analysis of *cis*-elements only	
Method: RT-PCR and restriction digest	Very cost effective and quick	Not very sensitive, only certain sites can be studied	Farré and Araya (2001), Staudinger et al. (2005)
Method: RT-PCR and sequence analysis	Semi-quantitative. Can study many sites per molecule	Not very sensitive, ratios of T to C tracks detect only high levels of editing	Farré and Araya (2001)
In vitro	Allows to change reaction conditions to study *cis*- as well as *trans*-elements (proteins)	Not all sites are edited at detectable levels	
Method: detection of U from labelled C	Sensitive and semi-quantitative	Expensive and labour-intensive to label specific C, can study only one site per molecule or all	Araya et al. (1992); applied to plastids: Hirose and Sugiura (2001)
Method: TDG	Sensitive and semi-quantitative. Can study several sites per molecule	Complex, labour-intensive procedure	Takenaka and Brennicke (2003)
Method: poisoned primer (not yet applied to mitochondria)	Relatively fast and easy. Semi-quantitative	Can study only one site per molecule. Many sites in close proximity could disrupt poisoned primer binding	Applied to plastids: Hegemann et al. (2005)

nucleotides in the upstream region and six nucleotides downstream are sufficient to confer the information to determine the nucleotide to be edited (Farré et al. 2001; Choury et al. 2004). As examples, the editing sites number 77 and 259 in the *cox2* mRNA were found to be correctly recognized when this pattern of specific nucleotides is maintained. Further distant nucleotide identities seem to have little influence, since relocating the 23 respective nucleotides to other positions in the mRNA still results in editing at the correct C nucleotide (Choury et al. 2004).

While the maximal requirement of 23 nucleotides is similar for these two sites, the importance of individual nucleotide identities such as the nucleotide immediately 3′ of the edited C differ between the two loci. This observation was interpreted to indicate that the various editing sites in plant mitochondria are recognized by different *trans*-recognition factors in editosome assembly.

The establishment of an in organello editing system for mitochondria from maize shows that, in some instances, more than the immediate vicinity of an editing site may be required for efficient editing in organello (Staudinger and Kempken 2003; Staudinger et al. 2005). While transcripts synthesized in the purified mitochondria from an introduced *cox2* gene from the rather distant dicot *Arabidopsis thaliana* were edited at several sites, an *atp6* mRNA from the more closely related plant *Sorghum bicolor* was not edited at all, although in these transcripts all editing sites are identical with maize and their respective surrounding sequence are also largely conserved. These results suggest that variations in distant regions, possibly even the structures of 5′- and/or 3′-untranslated sequences, might influence RNA editing at some sites.

These findings obtained from elegant in organello analysis are complemented and sustained by investigations with in vitro lysates from plant mitochondria.

6.2   *In Vitro RNA Editing*

The recently established in vitro RNA editing systems from pea and cauliflower mitochondria showed that, for recognition of the first editing site in the *atp9* mRNA by the RNA editing activity, about 20 nucleotides are essential upstream, 40 are optimal, and basically none are necessary downstream (Fig. 2). Analysis of the *cis-*

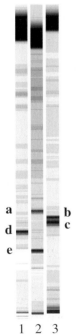

Fig. 2 In vitro analysis of RNA editing in plant mitochondria by the TDG-mismatch analysis method. RNA templates were incubated with mitochondrial extracts from cauliflower inflorescences, and the RT-PCR products were analysed for C to T changes. A sample gel is shown in which different editing sites were investigated. Track *1* shows the first *atp9* editing site (*d*), track *2* the same site (*e*; gel migration differs from track *1* because the template RNAs differ) together with the next downstream site (*a*), which is separated by 30 nucleotides. Track *3* shows two closely spaced editing sites (*b* and *c*), which are separated by two nucleotides in the *atp4* mRNA template. The strong signal at the top of each lane is emitted by the unedited and, thus, uncut template molecules. The procedure has been described in detail in Takenaka and Brennicke (2003)

requirements by targeted mutations of the template and competition experiments narrowed the sequence requirements for the site specificity to the region 5 to 20 nucleotides upstream of this site in the *atp9* mRNA (Takenaka et al. 2004). The region up to 40 nucleotides upstream strongly influences the efficiency of in vitro editing (Neuwirt et al. 2005). If this sequence is deleted, then the editing reaction still proceeds but at less than 50% of the activity recorded with the full-length template. This region surprisingly also influences editing at the next downstream site, which is 30 nucleotides 3′ of the first site, across a total distance of about 70 nucleotides (van der Merwe et al. 2006). For this site, the native cauliflower sequence enhances the reaction whereas the heterologous pea sequence in the cauliflower lysate inhibits editing.

An analogous observation of a long-distance effect has been observed and analysed by introducing the heterologous sorghum *atp6* sequence into maize mitochondria by electroporation (Staudinger et al. 2005). Only when most of the sequence upstream of the conserved *atp9* core reading frame of sorghum origin has been replaced by the homologous maize sequence is any editing observed at all. This suggests that there is some RNA editing entry motif in this upstream region, which is species-specific and acts for the entire length of the transcript. This observation indirectly suggests a scanning mode of the progression of the 'editosome' along the mRNA molecule which requires a specific engagement sequence to allow any binding to the 5′ region at all. Conserved nucleotide sequence motifs have indeed been described upstream of open reading frames in plant mitochondria (Schuster and Brennicke 1989; Pring et al. 1992), and these could possibly be involved in such a function.

Rather than a straightforward scanning progress of the entire editosome, certain species-specific sequence motifs might be required to engage the RNA helicase and potential cofactors, which need to clear the template RNA for access of the editosome. A sequence motif active in one species may not be sufficient, or even a hindrance, for a helicase (complex) from another plant species. The resulting mode of action would resemble an intermediate between a full-fledged scanning and a pure hit-and-run mechanism, and could still explain the apparent randomness of partially edited sites in the in vivo mRNA population.

Although the enhancing action of such common and more distant sequences can affect two or more sites in a given RNA template, the generally emerging picture about site specificity is that each site carries its own recognition context in its vicinity. Various in vitro competition experiments and the results of the transfection assays with isolated wheat mitochondria lead to this conclusion (Choury et al. 2004; Verbitskiy et al. 2006). The competition assays showed that only a homologous sequence competes in editing at a monitored site, while excess competition with the sequence contexts of different editing sites have little influence on the reaction.

In addition to these specific upstream elements, individual site-specific nucleotide positions downstream have been found to be crucial for efficient editing. These sequence requirements indicate that individual recognition elements vary between different editing sites.

The two experimental approaches, in vitro assays with dicots and in organello experiments with monocots, thus suggest that while between plants the recognition parameters of a given RNA editing site may be similar, they can vary between individual sites. Therefore, distinct *trans*-factors must be involved which identify individual RNA editing sites. For the large number of 400 to 500 editing events in plants mitochondria, a family of several hundred distinct *trans*-factors has to be postulated.

6.3   Trans-Factors – Proteins only?

Most of the 400 RNA editing sites in flowering plant mitochondria are found in mRNAs. Consequently, the sequence vicinities of homologous sites are highly conserved between different species and are presumably recognized by likewise conserved *trans*-factors.

Given the sheer number of editing sites which needs to be specified in mitochondrial transcripts, a correspondingly large number of *trans*-factors will be required. If sequence-specific recognition is mediated by proteins as *trans*-factors alone, as in the apoB editing, then we have to expect comparable numbers of different site-specific proteins, i.e. genes in the nuclear genome of plants.

A few years ago, one such protein (and corresponding gene) family has indeed been identified, and which encodes the family of so-called PPR proteins, aptly named for the presence of numerous pentatricopeptide repeats. In plants, this class of proteins has been found to be highly amplified, encompassing several hundred members (Small and Peeters 2000; Lurin et al. 2004). Prediction programs suggest that a large proportion of the encoded proteins can be targeted to the organelles, and several N-terminal regions have been shown experimentally in transgenic plants to be competent in bringing fused reporter proteins to either plastids or mitochondria, or both (Lurin et al. 2004).

Theoretical considerations suggest that these proteins could be nucleic acid binding. Based on genetic and biochemical evidence, it was found that individual proteins are required for various processing steps of RNAs in plastids. The first clear connection to RNA editing was obtained with a mutant of *Arabidopsis thaliana*, in which mutation of one PPR encoding gene results in the loss of one specific RNA editing site in the chloroplast (Kotera et al. 2005). By extrapolation, the family of PPR proteins is therefore the most likely contributor of the site-specific recognition factors in RNA editing in plastids as well as in mitochondria.

Alternatively or additionally, RNA sequences might help to guide the (as yet hypothetical) editosome to specific sites. Such a guide function may be provided by *cis*- or *trans*-acting RNA molecules but cannot reside in the sequence vicinity of the editing sites alone, since no common sequence motifs or related groups of such can be identified around the different C to U conversion sites. In plant mitochondria, there are ample space and numerous sequence duplications and similarities for such *trans*-acting guide RNAs to be easily encoded by the mito-chondrial DNA itself. However, no direct and conclusive evidence for the involve-ment of guide RNAs has yet been published and, with the identification of the family of PPR proteins, numerous proteins with the potential to bind RNA by sequence-specific recognition are available, so that per se additional RNA does not need to be invoked.

7 Future Tasks in the Analysis of RNA Editing in Plant Mitochondria

The most pressing question not resolved in the 15 years since RNA editing in plant mitochondria has been recognized is the open problem of the enzyme(s) involved and the underlying biochemistry (Fig. 3). How is the C to U alteration actually done in the RNA? Is it a straightforward deamination reaction? Or is the biochemical process a transamination to large acceptor molecules? Could such acceptor mole-cules be unspecific or dedicated RNA molecules?

With the recognition of the PPR protein family, the question of the specificity of RNA editing seems to have come to a realistic solution. These proteins are the most likely candidates for the specificity-mediating *trans*-factors but their individual assignment to given RNA editing sites still needs to be shown experimentally. How do these proteins interact with the enzymatic moiety? Is there a separate enzyme at all or do the PPR proteins each carry a domain competent in deamination of the polymerized C to a U?

The nature and structure of the components of the postulated 'editosome' need to be clarified (Fig. 3). What is the function of the potentially involved helicase? What is the role of unspecific RNA binding proteins such as the RBP proteins, which are rather abundant in plant mitochondria and presumably can attach to any

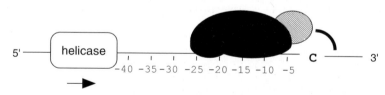

Fig. 3 Cartoon of our current model of RNA editing in plant mitochondria. The edited C is indi-cated in the *horizontal line* representing the RNA template molecule. The *black shape* symbolizes the specificity factor which recognizes the sequence region located between 5 (or 10) to about 20–25 nucleotides upstream of the edited C. The *hatched circle* represents the as yet hypothetical editing enzyme, which performs the actual reaction of deaminating the C to a U. This reaction may be influenced by the identity of the adjacent downstream nucleotide(s). The potentially involved RNA helicase attaches to the template RNA and makes this molecule accessible for the actual RNA editing activity by resolving secondary structures and/or by removing unspecific RNA bind-ing proteins

RNA (Vermel et al. 2002)? Answers to these questions may ultimately allow us to assemble a model for the RNA maturation steps in plant mitochondria, and their order and location in the mitochondrial organelle. We may then be able to understand how far transcription and RNA maturation are coupled and channelled within the organelle to yield regulated amounts of the required protein products.

Acknowledgements Mizuki Takenaka was a fellow of the JSPS and Johannes Andreas van der Merwe was a fellow of the DAAD. Work in the authors' laboratory is supported by grants from the Deutsche Forschungsgemeinschaft.

References

Araya A, Domec C, Begu D, Litvak S (1992) An in vitro system for the editing of ATP synthase subunit 9 mRNA using wheat mitochondrial extracts. Proc Natl Acad Sci USA 89:1040–1044

Benne R, Van den Burg J, Brakenhoff JPJ, Sloof P, Van Boom JH, Tromp MC (1986) Major transcript of the frameshifted coxII gene from trypanosome mitochondria contains four nucleotides not encoded in the DNA. Cell 46:819–826

Bock R, Herrmann M, Kössel H (1996) *In vivo* dissection of *cis*-acting determinants for plastid RNA editing. EMBO J 15:5052–5059

Bock R, Herrmann M, Fuchs M (1997) Identification of critical nucleotide positions for plastid RNA editing-site recognition. RNA 3:1194–1200

Carillo C, Bonen L (1997) RNA editing status of nad7 intron domains in wheat mitochondria. Nucleic Acids Res 25:403–409

Carter CW (1998) Nucleoside deaminases for cytidine and adenosine: comparisons with deaminases acting on RNA. In: Grosjean H, Benne R (eds) Modification and editing of RNA. ASM Press, Washington, DC, pp 363–376

Chang BH-J, Lau PP, Chan L (1998) Apolipoprotein B mRNA editing. In: Grosjean H, Benne R (eds) Modification and editing of RNA. ASM Press, Washington, DC, pp 325–342

Chateigner-Boutin A-L, Hanson MR (2002) Cross-competition in transgenic chloroplasts expressing single editing sites reveals shared cis elements. Mol Cell Biol 22:8448–8456

Chaudhuri S, Maliga P (1996) Sequences directing C to U editing of the plastid psbL mRNA are located within a 22 nucleotide segment spanning the editing site. EMBO J 15:5958–5964

Chaudhuri S, Carrer H, Maliga P (1995) Site-specific factor involved in the editing of the psbL mRNA in tobacco plastid. EMBO J 14:25951–25957

Choury D, Farré J-C, Jordana X, Araya A (2004) Different patterns in the recognition of editing sites in plant mitochondria. Nucleic Acids Res 32:6397–6406

Covello PS, Gray MW (1989) RNA editing in plant mitochondria. Nature 341:662–666

Covello PS, Gray MW (1993) On the evolution of RNA editing. Trends Genet 9:265–268

Fairman ME, Maroney PA, Wang W, Bowers HA, Gollnick P, Nilsen TW, Jankowsky E (2004) Protein displacement by DexH/D "RNA helicases" without duplex unwinding. Science 304:730–734

Faivre-Nitschke SE, Grienenberger JM, Gualberto JM (1999) A prokaryotic-type cytidine deaminase from *Arabidopsis thaliana*: gene expression and functional characterization. Eur J Biochem 263:896–903

Farré J-C, Araya A (2001) Gene expression in isolated plant mitochondria: high fidelity of transcription, splicing and editing of a transgene product in electroporated organelles. Nucleic Acids Res 29:2484–2491

Farré J-C, Leon G, Jordana X, Araya A (2001) Cis recognition elements in plant mitochondrion RNA editing. Mol Cell Biol 21:6731–6737

Giegé P, Brennicke A (1999) RNA editing in Arabidopsis mitochondria effects 441 C to U changes in ORFs. Proc Natl Acad Sci USA 96:15324–15329

Gott JM, Visomirski-Robic LM (1998) RNA editing in *Physarum* mitochondria. In: Grosjean H, Benne R (eds) Modification and editing of RNA. ASM Press, Washington, DC, pp 395–412

Greeve J, Navaratnam N, Scott J (1991) Characterization of the apolipoprotein B mRNA editing enzyme: no similarity to the proposed mechanism of RNA editing in kinetoplastid protozoa. Nucleic Acids Res 19:3569–3576

Groth-Malonek M, Knoop V (2005) Bryophytes and other basal land plants: the mitochondrial perspective. Taxon 54:293–297

Gualberto JM, Lamattina L, Bonnard G, Weil J-H, Grienenberger JM (1989) RNA editing in wheat mitochondria results in the conservation of protein sequences. Nature 341:660–662

Gualberto JM, Weil J-H, Grienenberger JM (1990) Editing of the wheat coxIII transcript: evidence for twelve C-to-U and one U-to-C conversions and for sequence similarities around editing sites. Nucleic Acids Res 18:3771–3776

Hajduk SL, Sabatini RS (1998) Mitochondrial mRNA editing in kinetoplastid protozoa. In: Grosjean H, Benne R (eds) Modification and editing of RNA. ASM Press, Washington, DC, pp 377–394

Handa H (2003) The complete nucleotide sequence and RNA editing content of the mitochondrial genome of rapeseed (*Brassica napus* L.): comparative analysis of the mitochondrial genomes of rapeseed and *Arabidopsis thaliana*. Nucleic Acids Res 31:5907–5916

Hegemann CE, Hayes ML, Hanson MR (2005) Substrate and cofactor requirements for RNA editing of chloroplast transcripts in *Arabidopsis* in vitro. Plant J 42:124–132

Hernould M, Suharsono S, Litvak S, Araya A, Mouras A (1993) Male-sterility induction in transgenic tobacco plants with an unedited *atp9* mitochondrial gene from wheat. Proc Natl Acad Sci USA 90:2370–2374

Hiesel R, Wissinger B, Schuster W, Brennicke A (1989) RNA editing in plant mitochondria. Science 246:1632–1634

Hirose T, Sugiura M (2001) Involvement of a site-specific trans-acting factor and a common RNA-binding protein in the editing of chloroplast RNA: development of an *in vitro* RNA editing system. EMBO J 20:1144–1152

Knoop V (2004) The mitochondrial DNA of plants: peculiarities in phylogenetic perspective. Curr Genet 46:123–139

Kotera E, Tasaka M, Shikanai T (2005) A pentatricopeptide repeat protein is essential for RNA editing in chloroplasts. Nature 433:326–330

Kubo N, Kadowaki K (1997) Involvement of 5′ flanking sequence for specifying RNA editing sites in plant mitochondria. FEBS Lett 413:40–44

Lippok B, Wissinger B, Brennicke A (1994) Differential RNA editing in closely related introns in *Oenothera* mitochondria. Mol Gen Genet 243:39–46

Lurin C, Andrés C, Aubourg S, Bellaoui M, Bitton F, Bruyère C, Caboche M, Debast C, Gualberto J, Hoffmann B, Lecharny A, Le Ret M, Martin-Magniette M-L, Mireau H, Peeters N, Renou J-P, Szurek B, Taconnat L, Small I (2004) Genome-wide analysis of *Arabidopsis* pentatricopeptide repeat proteins reveals their essential role in organelle biogenesis. Plant Cell 16:2089–2103

Mahendran R, Spottswood MR, Miller DL (1991) RNA editing by cytidine insertion in mitochondria of *Physarum polycephalum*. Nature 349:434–438

Malek O, Lättig K, Hiesel R, Brennicke A, Knoop V (1996) RNA editing in bryophytes and a molecular phylogeny of land plants. EMBO J 15:1403–1411

Marchfelder A, Binder S (2004) Plastid and plant mitochondrial RNA processing and RNA stability. In: Daniell H, Chase CD (eds) Molecular biology and biotechnology of plant organelles. Springer, Dordrecht, pp 261–294

Miyamoto T, Obokata J, Sugiura M (2002) Recognition of RNA editing sites is directed by unique proteins in chloroplasts: biochemical identification of cis-acting elements and trans-acting factors involved in RNA editing in tobacco and pea chloroplasts. Mol Cell Biol 22:6726–6734

Miyamoto T, Obokata J, Sugiura M (2003) A site-specific factor interacts directly with its cognate RNA editing site in chloroplast transcripts. Proc Natl Acad Sci USA 101:48–52

Mulligan M (2004) RNA editing in plant organelles. In: Daniell H, Chase CD (eds) Molecular biology and biotechnology of plant organelles. Springer, Dordrecht, pp 239–260

Navaratnam N, Fujino T, Bayliss J, Jarmuz A, How A, Richardson N, Somasekaram A, Bhattacharya S, Carter C, Scott J (1998) Escherichia coli cytidine deaminase provides a molecular model for ApoB RNA editing and a mechanism for RNA substrate recognition. J Mol Biol 275:695–714

Neuwirt J, Takenaka M, van der Merwe JA, Brennicke A (2005) An in vitro RNA editing system from cauliflower mitochondria: editing site recognition can vary between different plant species. RNA 11:1563–1570

Notsu Y, Masood S, Nishikawa T. Kubo N, Akiduki G, Nakazono M, Hirai A, Kadowaki K (2002) The complete sequence of the rice (*Oryza sativa* L.) mitochondrial genome: frequent DNA acquisition and loss during the evolution of flowering plants. Mol Genet Genomics 268:434–445

Oda K, Yamato K, Ohta E, Nakamura Y, Takemura M, Nozato N, Akashi K, Kanegae T, Ogura Y, Kohchi T, Ohyama K (1992) Gene organization deduced from the complete sequence of liverwort *Marchantia polymorpha* mitochondrial DNA: a primitive form of plant mitochondrial genome. J Mol Biol 223:1–7

Phreaner CG, Williams MA, Mulligan RM (1996) Incomplete editing of rps12 transcripts results in the synthesis of polymorphic polypeptides in plant mitochondria. Plant Cell 8:107–117

Pring DR, Mullen JA, Kempken F (1992) Conserved sequence blocks 5′ to start codons of plant mitochondrial genes. Plant Mol Biol 19:313–317

Rajasekhar VK, Mulligan RM (1993) RNA editing in plant mitochondria: [alpha]-phosphate is retained during C-to-U conversion in mRNAs. Plant Cell 5:1843–1852

Reed ML, Peeters NM, Hanson MR (2001) A single alteration 20 nt 5′ to an editing target inhibits chloroplast RNA editing in vivo. Nucleic Acids Res 29:1507–1513

Schuster W, Brennicke A (1989) Conserved sequence elements at putative processing sites in plant mitochondria. Curr Genet 14:187–192

Schuster W, Hiesel R, Wissinger B, Brennicke A (1990) RNA editing in the cytochrome b locus of the higher plant *Oenothera* includes a U to C transition. Mol Cell Biol 10:2428–2431

Small I, Peeters N (2000) The PPR motif – A TPR-related motif prevalent in plant organellar proteins. Trends Biochem Sci 25:46–47

Staudinger M, Kempken F (2003) Electroporation of isolated higher-plant mitochondria: transcripts of an introduced *cox2* gene, but not an atp6 gene, are edited *in organello*. Mol Genet Genomics 269:553–561

Staudinger M, Bolle N, Kempken F (2005) Mitochondrial electroporation and in organello RNA editing of chimeric *atp6* transcripts. Mol Genet Genomics 273:130–136

Takenaka M, Brennicke A (2003) *In vitro* RNA editing in pea mitochondria requires NTP or dNTP, suggesting involvement of an RNA helicase. J Biol Chem 278:47526–47533

Takenaka M, Neuwirt J, Brennicke A (2004) Complex *cis*–elements determine an RNA editing site in pea mitochondria. Nucleic Acids Res 32:4137–4144

Turmel M, Otis C, Lemieux C (2003)The mitochondrial genome of *Chara vulgaris*: insights into the mitochondrial DNA architecture of the last common ancestor of green algae and land plants. Plant Cell 15:1888–1903

van der Merwe JA, Takenaka M, Neuwirt J, Verbitskiy D, Brennicke A (2006) RNA editing sites in plant mitochondria can share cis-elements. FEBS Lett 580:268–272

Verbitskiy D, Takenaka M, Neuwirt J, van der Merwe JA, Brennicke A (2006) Partially edited RNAs are intermediates of RNA editing in plant mitochondria. Plant J 47:408–416

Vermel M, Guermann B, Delage L, Grienenberger JM, Marechal-Drouard L, Gualberto JM (2002) A family of RRM-type RNA-binding proteins specific to plant mitochondria. Proc Natl Acad Sci USA 99:5866–5871

Yu W, Schuster W (1995) Evidence for a site-specific cytidine deamination reaction involved in C to U editing in plant mitochondria. J Biol Chem 270:18227–18233

Zabaleta E, Mouras A, Hernould M, Suharsono S, Araya A (1996) Transgenic male-sterile plant induced by an unedited *atp9* gene is restored to fertility by inhibiting its expression with antisense RNA. Proc Natl Acad Sci USA 93:11259–11263

RNA Editing in Chloroplasts

Masahiro Sugiura

> *"Amino acid sequences as well as open reading frames from chloroplast DNA sequences cannot be deduced from the straightforward rules of the genetic code."*
>
> *Hans Kössel*

Abstract RNA editing has been observed in chloroplast transcripts from most land plants, but not from algae. In higher plant chloroplasts, C-to-U conversion occurs at around 30 specific sites in mRNAs, many of which result in amino acid alteration. Occasionally, RNA editing creates an AUG initiation codon from ACG, and a UAA termination codon from CAA. No obvious consensus element or secondary structure was found in the sequences surrounding these editing sites. Chloroplast transformation approaches have revealed that *cis*-elements for editing are commonly located in the upstream and proximal regions of editing sites. Development of an in vitro chloroplast RNA editing system has enabled us to investigate biochemical processes of RNA editing, e.g., detection of *trans*-acting factors.

1 Introduction

Chloroplasts are plant-specific organelles that contain the entire machinery for photosynthesis. They also are involved in the biosynthesis of amino acids, nucleotides, lipids, starch, and so on. In addition, chloroplasts possess their own

Graduate School of Natural Sciences, Nagoya City University, Nagoya 467-8501, Japan; sugiura@nsc.nagoya-cu.ac.jp genetic system, and its genome is represented genetically as a circular form of double-stranded DNA. The chloroplast genome of higher plants is about 150 kbp in length, and contains around 80 protein-coding genes and 35 genes encoding stable RNA species, rRNAs, and tRNAs (Sugiura 1992; Wakasugi et al. 2001). Expression of chloroplast genes is controlled during development, and in response to environmental cues. In the chloroplast of higher plants, many genes are transcribed with two types of RNA polymerases, one similar to *E. coli* RNA polymerase, and another related to phage T3/T7 RNA polymerase, producing generally polycistronic precursors that are processed into complex sets of overlapping transcripts including monocistronic mRNAs. During this process, some of the transcripts are spliced. Translation in chloroplasts is in many aspects unique, and requires usually monocistronic mRNAs. These RNA processing and translation steps have been recognized as important regulatory steps in chloroplast gene expression.

RNA editing in chloroplasts was first reported in a maize transcript (Hoch et al. 1991). The ribosomal protein L2 genes (*rpl2*) from maize and rice are known to have an ACG triplet at the position corresponding to the ATG initiation codon in other plant *rpl2* genes analyzed. Analysis of the maize *rpl2* mRNA revealed that the ACG is converted into a canonical AUG initiation codon by a C-to-U base change. RNA editing is not limited to initiation codons, but has been observed at internal codons (e.g., Maier et al. 1992) and sometimes in untranslated regions (UTRs). RNA editing has been found in chloroplast transcripts from all major lineages of land plants (Freyer et al. 1997), but has not been reported in those of algae and cell transcripts from cyanobacteria. Most editing events found so far in chloroplasts are C-to-U conversions, but U-to-C inverse editing, in addition to C-to-U changes, has been reported for the chloroplast transcripts from a bryophyte, the hornwort *Anthoceros formosae* (Yoshinaga et al. 1996; Duff and Moore 2005). However, no editing was reported another bryophyte, the liverwort *Marchantia polymorpha*, whereas another liverwort, *Bazzania trilobata*, exhibits RNA editing at least in *ndhB* and *rbcL* transcripts (Freyer et al. 1997). RNA editing in chloroplasts has been reviewed (as of 2000) by Bock (2000, 2001), Tsudzuki et al. (2001), Mulligan (2004), and Shikanai (2006).

2   RNA Editing Sites

Nucleotide sequences in chloroplast DNAs are often modified at the level of RNA, by RNA editing as well as by RNA splicing. Therefore, identification of editing sites in transcripts is essential for understanding precise coding information in the chloroplast genome, and systematic search for editing sites has been made in chloroplast transcripts from several plant species. Candidate editing sites were first screened by aligning all predicted protein sequences and their coding gene sequences from a plant species with those from other plants, and from *Marchantia polymerpha*, algae, *Escherichia coli* K12 and cyanobacteria, which have so far reported no RNA editing sites. Among amino acid residues deviated from conserved residues in other species, C-to-U exchanges at the RNA level potentially restoring the conserved ones were then tested experimentally by sequencing corresponding transcripts (cDNA). Systematic search identified 27 editing sites in maize (Maier et al. 1995; Bock et al. 1997a), 26 in black pine (Wakasugi et al. 1996), 38 in tobacco (Hirose et al. 1999; Sasaki et al. 2003, 2006; Kahlau et al. 2006), 35 in *Nicotiana sylvestris* (Sasaki et al. 2003; Sasaki 2005), 34 in *Nicotiana tomentosiformis* (Sasaki et al. 2003; Sasaki 2005), 31 in *Atropa belladonna* (deadly nightshade, Schmitz-Linneweber et al. 2002), 26 in rice (Corneille et al. 2000; Inada et al. 2004), 23 in sugarcane (Calsa et al. 2004), 28 in *Arabidopsis* (Lutz and Maliga 2001; Tillich et al. 2005), 36 in tomato (Kahlau et al. 2006), and 44 in *Phalaemopsis aphrodite* (Zeng et al. 2007). All these editing events are C-to-U conversions, and no U-to-C inverse editing was reported in these species. According to the present screen for potential editing sites, it is common that edited codons restore amino acids that are conserved in the corresponding proteins of other plants. Exceptions to this general trend are an RNA editing event observed at the third position of a serine codon (ucC to ucU) in the tobacco *atpA* transcript, not leading to amino acid change (silent editing, Hirose et al. 1996). Editing was also detected in the 5'UTRs of maize and rice *ndhG* mRNAs (Bock et al. 1997a; Corneille et al. 2000), and of Ginkgo *psbJ* transcripts (Kudla and Bock 1999). The pea *ndhG* site 1 (serine to leucine) was found by chance, because this site is conserved (serine or phenylalanine) and hence editing at this site had not been predicted (Inada et al. 2004); this is also the case for tobacco (Sasaki et al. 2006). *Arabidopsis* also has the *ndhG* site 1, and another unexpected site *ndhB* site 11 (serine to leucine) where serine is already conserved (Tillich et al. 2005). Therefore, some editing events lead to the diversification of the evolutionarily conserved amino acid sequence.

Table 1 lists editing sites so far reported from five angiosperm chloroplasts. Of the total of 64 different sites listed, only three (*ndhB* sites 3 and 8, and *rps14* site 1) are common between the five angiosperms, while 18 sites are unique to one species. These data raise the possibility that many editing sites were acquired in the evolution of angiosperms. No editing has so far been observed in tRNA and rRNA transcripts. As the current protocol to predict a potential editing site has its limitations, it is likely that additional editing sites are present in the transcripts, especially in non-coding regions, from higher plant chloroplasts.

The most unusual case is that of the hornwort *Anthoceros formosae* (Kugita et al. 2003). In its chloroplasts, a total of 509 C-to-U and 433 U-to-C conversions were identified in the transcripts of 68 genes and eight ORFs. The frequency of editing events is higher than that observed in plant mitochondria (441 sites in *Arabidopsis*, Giege and Brennicke 1999; see the chapter by Takenaka et al., this volume). RNA editing was also detected in one tRNA, and at three sites of an intron. Similarly, 350 editing sites were detected in the transcript from another fern, *Adiantum capillus-veneris* (Wolf et al. 2004). Hence, the editing pattern is completely different from those of angiosperm plants, and the mechanism of editing in these species is likely to be unique. A model for the evolution of plant organelle RNA editing was recently proposed by Tillich et al. (2006b).

Table 1 RNA editing sites in chloroplast transcripts[a]

Gene	Site	Position	Conversion	Tobacco	*Arabidopsis*	Pea	Rice	Maize
accD	1	(267)	S(uCg)→L(uUg)	t	+	+	t	n.a.
atpA	1	264	P(cCc)→L(cUc)	+	t	+	t	t
:	2	265	S(ucC)→S(ucU)	Δ	−	−	−	−
:	3	383	S(uCa)→L(uUa)	t	t	t	+	+
atpF	1	31	P(cCa)→L(cUa)	+	+	t	t	t
clpP	1	(187)	H(Cau)→Y(Uau)	n.a.	+	n.a.	n.a.	n.a.
matK	1	420	H(Cau)→Y(Uau)	t	t	t	+	+
:	2	(236)	H(Cau)→Y(Uau)	n.a.	+	n.a.	n.a.	n.a.
ndhA	1	17	S(uCg)→L(uUg)	t	t	t	t	+
:	2	114	S(uCa)→L(uUa)	+	−[b]	Δ	t	t
:	3	159	P(cCa)→L(uUa)	t	t	t	+	+
:	4	189	S(uCa)→L(uUa)	t	t	t	+	+
:	5	358	S(uCc)→F(uUc)	+	t	−	+	+
ndhB	1	50	S(uCa)→L(uUa)	+	+	+	t	t
:	2	156	P(cCa)→L(cUa)	+	+	t	+	+
:	3	196	H(Cau)→Y(Uau)	+	+	+	+	+
:	4	204	S(uCa)→L(uUa)	+	−	−	Δ	+
:	5	235	S(uCc)→F(uUc)	t	t	t	+	t
:	6	246	P(cCa)→L(cUa)	+	t	+	+	+
:	7	249	S(uCu)→F(uUu)	+	+	+	t	t
:	8	277	S(uCa)→L(uUa)	+	+	+	+	+
:	9	279	S(uCa)→L(uUa)	+	+	+	+	t
:	10	494	P(cCa)→L(cUa)	+	+	−	Δ	+
:	11	(291)	S(uCa)→L(uUa)	n.a.	+	n.a.	n.a.	n.a.
:	12	419	H(Cau)→Y(Uau)	t	+	t	t	t
ndhD	1	1	T(aCg)→M(aUg)	Δ	Δ	Δ	t	t
:	2	128	S(uCa)→L(uUa)	+	+	+[c]	t	t
:	3	293	S(uCa)→L(uUa)	t	+	−	+	+
:	4	225	S(uCg)→L(uUg)	+	+	+	t	t
:	5	(296)	P(cCc)→L(cUc)	t	+	t	t	t
:	6	200	S(uCa)→L(uUa)	Δ	t	t	t	t
:	7	433	S(uCa)→L(uUa)	+	t	n.a.	t	t
:	8	437	S(uCa)→L(uUa)	+	t	n.a.	t	t
ndhF	1	21	S(uCa)→L(uUa)	t	t	t	+	+
:	2	97	S(uCa)→L(uUa)	+	+	Δ	t	t
ndhG	−1	−10 nt	C→U	t	t	t	+	+
:	nd1	17	S(uCg)→L(uUg)	+	+	+	t	t
:	2	116	S(uCa)→L(uUa)	+	t	+	+[c]	t
petB	1	204	P(cCa)→L(cUa)	+	t	+	t	n.a.
:	2	(223)	P(cCa)→L(cUa)	n.a.	n.a.	n.a.	−	+
petL	1	2	P(cCc)→L(cUc)	t	+	Δ	t	t
psaI	1	27	H(Cau)→Y(Uau)	−	n.a.	Δ	t	t
psbE	1	72	P(Cca)→S(Uca)	+	+	t	t	t
psbF	1	26	S(uCu)→F(uUu)	t	+	+	t	n.a.
psbL	1	1	T(aCg)→M(aUg)	+	t	t	t	t
psbZ	1	17	S(uCc)→F(uUc)	t	n.a.	+	−	n.a.

rpl2	1	1	T(aCg)→M(aUg)	t	t	t	Δ	+
rpl20	1	103	S(uCa)→L(uUa)	+	t	t	t	+
rpoA	1	277	S(uCa)→L(uUa)	Δ	t	t	t	t
rpoB	1	113	S(uCu)→F(uUu)	+	+	t	t	t
:	2	158	S(uCa)→L(uUa)	+	t	t	Δ	+
:	3	184	S(uCa)→L(uUa)	+	+	Δ	Δ	+
:	4	189	S(uCa)→L(uUa)	t	t	+	Δ	+
:	5	208	P(cCg)→L(cUg)	t	t	t	−	+
:	6	667	S(uCu)→F(uUu)	+	t	+	t	t
:	7	(811)	S(uCa)→L(uUa)	n.a.	+	n.a.	n.a.	n.a.
rpoC1	1	21	S(uCa)→L(uUa)	+	t	Δ	t	t
rpoC2	1	778	S(uCg)→L(uUg)	t	t	n.a.	t	+
:	2	1,248	S(uCa)→L(uUa)	+	t	t	+	t
rps2	1	83	S(uCa)→L(uUa)	+	t	+	t	t
:	2	45	T(aCa)→I(aUa)	+	n.a.	+	+	n.a.
rps8	1	61	S(uCa)→L(uUa)	t	t	t	+	+
rps14	1	27	S(uCa)→L(uUa)	+	+	+[c]	+	+
:	2	50	P(cCa)→L(cUa)	+	+	t	t	t
ycf3	1	15	S(uCc)→F(uUc)	t	[d]	n.a.	t	+
:	2	62	T(aCg)→M(aUg)	t	t	t	Δ	+

[a]Position in codon is given with respect to the tobacco initiation codon (parentheses are from other species). +, Edited, t, no editing as T in DNA, −, no editing detected as C in DNA, n.a., not analyzed, Δ, partial editing. Tobacco sites are from Hirose et al. (1999), Sasaki et al. (2003, 2006), and Sasaki (2005), and Kahlau et al. (2006); *Arabidopsis* sites from Lutz and Maliga (2001), and Tillich et al. (2005a); pea sites from Inada et al. (2004); maize sites from Maier et al. (1995), Bock et al. (1997a), and Peeters and Hanson (2002); rice sites from Corneille et al. (2000), and Inada et al. (2004)

[b]One group reported editing

[c]P(cCa)→L(cUa) editing

[d]Gap

3 Functional Aspects of RNA Editing

The functional significance of RNA editing has been demonstrated using chloroplast transformation technologies (Svab and Maliga 1993; Bock 2001). The *psbF* mRNA is edited in spinach, changing a serine into a conserved phenylalanine codon, while the corresponding site in tobacco is a phenylalanine codon (thus, no editing). The spinach *psbF* editing site introduced into the tobacco *psbF* gene remained unedited and led to a phenotype of photosynthetic mutant, and hence the editing was essential for protein function (Bock et al. 1994). Editing of the tobacco *petB* site (proline to leucine) inserted into *Chlamydomonas* chloroplasts did not occur, and the alga is non-photosynthetic; hence, a leucine residue at this position is required for cytochrome b6 function (Zito et al. 1997). In peas, the *accD* mRNA encoding the β-polypeptide of acetyl-CoA carboxylase is edited to convert the serine to leucine codon, and this editing was shown to be required for functional enzyme activity by comparing the unedited and edited recombinant enzymes (Sasaki et al. 2001). This was the first biochemical evidence that editing is essential for a functional protein.

RNA editing sometimes creates AUG initiation codons from ACG triplets in *psbL* and *ndhD* transcripts, as was first reported in maize and rice *rpl2* mRNAs by Hans Kössel's group. It is possible that conversion of ACG into an AUG allows the mRNA to start translation. This hypothesis was confirmed using a chloroplast in vitro translation system – the mRNA with AUG is translated, but not its precursor with ACG (Hirose and Sugiura 1997) – and using in vivo chloroplast transformation (Lutz et al. 2006). However, it cannot be completely ruled out that ACG can be utilized as an initiation codon in vivo, because the leek *ndhD* mRNA was unedited (ACG) but the NDH-D polypeptide was present (Del Campo et al. 2002), and because unedited *ndhD* transcripts were also associated with polysomes in vivo (Zandueta-Criado and Book 2004). Two RNA editing events create an AUG codon and an in-frame UAA codon within the same transcript from black pine ORF 62b, leading to the formation of a new protein-coding region that corresponds to the *petL* gene (Wakasugi et al. 1996). The editing to create stop codons is most likely to be essential for protein function, although there is no experimental proof of this to date. In hornwort chloroplasts, all nonsense codons in 52 protein-coding genes and seven ORFs are removed in the transcripts by U-to-C conversions, and five initiation and three termination codons are created by C-to-U changes (Kugita et al. 2003); a similar editing pattern was observed in the fern *Adiantum* (Wolf et al. 2004). These editing events are again likely to be necessary for functional proteins.

Among the edited sites identified, only limited cases have been proven to be functionally significant. As described above, there is silent editing, and editing events that do not restore conserved amino acids. In the *psbZ* editing site, pea acquires a phenylalanine codon (uUc) from serine (uCc) by RNA editing, but no editing occurs at the corresponding site in rice (uCa); hence, rice maintains a serine residue at this position, suggesting that editing of this site is not essential for *psbZ* protein (Inada et al. 2004). Similarly, nine other sites among the 64 editing sites (Table 1) remained unedited in some species at C residues in the genome. RNA editing was observed in the *petL* transcripts of spinach and other plant chloroplasts, many species harboring two or three sites. Targeted inactivation of tobacco *petL* did not impair plant growth under a variety of conditions, suggesting that *petL* is not essential in higher plants and hence these editing events have no functional significance (Fiebig et al. 2004). Taken together, some of the editing events are suggested to be neutral, or to have little effect on protein function.

4   Mechanism of RNA Editing: In Vivo Analysis

In angiosperm chloroplasts, RNA editing occurs at specific C residues, approximately one in one thousand C residues in whole transcripts. Therefore, the C residues to be edited should be recognized extremely precisely and efficiently, and then the C residues can be converted into U residues. The technology for the genetic transformation of tobacco chloroplasts has facilitated the study of RNA editing in an in vivo system (Bock 2004). In tobacco chloroplasts, editing of an ACG triplet to AUG creates the initiation codon for *psbL* and *ndhD* transcripts. The *psbL* portion (98 bp) encompassing the editing site was inserted into a spacer region of the chloroplast genome. Expression of the inserted *psbL* fragment led to a significant decrease in editing of the endogenous *psbL* mRNA, but not of editing sites in other mRNAs, suggesting that site-specific factors are involved in editing, and that the factors are present in limiting amounts (Chaudhuri et al. 1995). Deletion and mutation analyses indicated that a 22-nucleotide (nt) segment is sufficient for editing, including 16 nt upstream and 5 nt downstream of the *psbL* editing site, and that the A residue in front of the C to be edited is essential for editing (Chaudhuri and Maliga 1996). As for *psbL*, editing of the *ndhD* site requires a depletable *trans*-factor, but distinct from that for *psbL* mRNA editing.

The *cis*-elements for the tobacco *ndhB* sites IV and V (separated by 8 nt) were found to reside within the 22-nt upstream and 22-nt downstream regions of sites IV and V, respectively; the sequence immediately upstream of site IV was identified to be essential for the editing of both sites, while the sequence downstream of site V had only a minor effect on the efficiency of editing at both sites (Bock et al. 1996). Further analysis with a scanning point mutagenesis on the 8-nt spacer between sites IV and V showed that only few nt positions are important in addition to the upstream sequence, and that these closely adjacent sites are edited independently, and not in a polar fashion (Bock et al. 1997b). Furthermore, deletion and insertion experiments showed that the distance from the upstream sequence element of the *ndhB* sites is critical for editing, suggesting that the editing machinery measures precisely this distance (Hermann and Bock 1999). The RNA editing sites I and IV present in maize *rpoB* mRNA were inserted into tobacco chloroplasts (Reed and Hanson 1997). Site I was edited in tobacco, and reduced editing of the endogenous site I (site 2 in tobacco), suggesting the presence of a depletable *trans*-factor. However, the maize site IV did not confer editing, suggesting the absence of a *trans*-factor for site IV in tobacco (T in the tobacco genome). Further analysis of the tobacco site 2 with point mutations demonstrated that the residue at position −20 5′ to site 2 is critical for editing in vivo, indicating that the *cis*-element for site 2 includes position −20 (Reed et al. 2001b). In vivo transgene assays using an edited *ndhF* gene (its editing site was converted into T) demonstrated that editing of the endogenous *ndhF* site was reduced. This suggested that a *trans*-factor can bind to edited mRNAs, and that the C to be edited is not critical for *trans*-factor recognition (Reed et al. 2001a).

The participation of site-specific *trans*-acting factors was also demonstrated by an interspecific protoplast fusion approach; editing at the spinach *psbF* editing site inserted into tobacco (see above) could be restored by fusing the tobacco cells with spinach cells, suggesting that the *trans*-factor is supplied from the spinach nucleo-cytoplasm (Bock and Koop 1997). Tobacco and deadly night-shade are incompatible species of the family Solanaceae, but readily produce hybrids and cybrids. Cybrids with a nuclear genome from deadly nightshade and a chloroplast genome from tobacco display an albino phenotype; the albino phenotype develops as a result of a defect in RNA editing of a tobacco *atpA* site, suggesting the absence of *trans*-factor for the site at no corresponding site in deadly nightshade (Schmitz-Linneweber et al. 2005). The extraplastidic origin of *trans*-factors was supported by the observations that chloroplast translation is not required for editing (Zeltz et al. 1993; Halter et al. 2004). However, chloroplast translation is required for a limited number of editing sites (Karcher and Book 1998), suggesting that all components of the chloroplast editing system are not of nuclear origin, at least for some specific sites, e.g., the *ndhB* site III.

A genetic approach was successfully applied for the finding of a nuclear gene involved in RNA editing in *Arabidopsis* chloroplasts. Using a screening system to identify impaired NADH dehydrogenase activity, a gene was isolated and found to encode the PPR protein (CRR4) of 606 amino acids with a predicted transit-peptide (Kotera et al. 2005). Sequence analysis of 20 editing sites from this mutant revealed that the *ndhD* site 1 (to create AUG) remained unedited. As PPR proteins are suggested to play important roles in organelle biogenesis, probably via binding to organelle transcripts (Lurin et al. 2004), the CRR4 protein was proved to interact specifically with the *ndhD* mRNA (Okuda et al. 2006). Similarly, another PPR protein (CRR21) was found to be involved in editing of *ndhD* site 2 (Okuda et al. 2007).

It was generally accepted that there is a single *trans*-factor for each chloroplast editing site. A comprehensive analysis of the editing efficiency of tobacco sites with highly expressed transgenes carrying editing sites revealed RNA editing sites can be grouped into clusters that carry conserved nucleotides in the upstream region, hence supporting a model in which the same *trans*-factor recognizes members of the same editing cluster (Chateigner-Boutin and Hanson 2002, 2003). Within the upstream region, sequence elements critical for efficient editing of tobacco *rpoB* mRNAs are also present in the *psbL* and *rps14* mRNAs (Hayes et al. 2006). When a heterogonous editing site was introduced into tobacco chloroplasts (no corresponding site in tobacco), no editing occurred, suggesting that tobacco lacks the corresponding *trans*-factor (see above). Apparently, editing sites and their cognate *trans*-factors co-evolved. However, an unexpected case has been reported of editing activity being present despite the absence of the target site. The *ndhA* mRNAs of barley, spinach and several others, but not tobacco, harbor an editing site close to the intron–3′exon junction; the editing at this site occurred after splicing, and probably the 5′exon includes at least part of the *cis*-element for editing (Del Campo et al. 2000; Schmitz-Linneweber et al. 2001). When spinach spliced *ndhA* gene was introduced into tobacco chloroplasts, the spinach site was edited in vivo in tobacco, and it was found that the editing activity in tobacco was derived from *N. tomentosiformis* (male progenitor; Schmitz-Linneweber et al. 2001). The maintenance of the editing activity without the corresponding site is puzzling, and it is speculated that the factor plays an additional role. Further analysis demonstrated that the *ndhA* editing activity is also present in *N. sylvestris*, the female progenitor of tobacco (Tillich et al. 2006a). By contrast, ten of the 64 editing sites (see above) remained unedited in some species at C in the genome, suggesting that the corresponding *trans*-factors have been lost in these species during evolution.

5   Mechanism of RNA Editing: In Vitro Analysis

Analyses using chloroplast transformation of tobacco have demonstrated the presence of *cis*-elements generally in front of editing sites, and of *trans*-acting factors. However, thorough analyses of sequence elements have to date generally not been performed, because of the time and labor required for producing a chloroplast transgenic tobacco, and in vivo techniques having their imitations for biochemical analyses of RNA editing reactions.

A long-awaited technical advance involving the development of an in vitro RNA editing system from tobacco chloroplasts has been achieved by Hirose and Sugiura (2001). The system supports accurate and efficient editing for a number of tobacco chloroplast mRNAs. A critical factor in developing an in vitro system is how to assay the activity of interest. Using the tobacco *psbL* mRNA as a model substrate, site-specific labeling of the mRNA at the editing site was used for the detection of the edited products (Fig. 1). The upstream and downstream parts of the mRNA (relative to the C residue to be edited) were synthesized separately. The 5' end of the downstream sequence, which constitutes the editing site, was labeled with ^{32}P using polynucleotide kinase, and ligated to the upstream part with T4 DNA ligase in the presence of a complementary bridge DNA oligonucleotide. The resulting *psbL* mRNA substrate was incubated with a chloroplast extract preparation. Intact chloroplasts were prepared from green tobacco leaves of 5–10 cm length. The isolated chloroplasts were lysed with 0.2% Triton X-100 in 2 M KCl, followed by centrifugation. The resulting supernatant fractions were extensively dialyzed. Extraction with a high concentration of salt is essential to obtain active fractions, suggesting that the editing machinery should be dissociated from super-complexes or membrane structures. After incubation, RNA was isolated, digested with nuclease P1, and separated by cellulose thin layer chromatography. Finally, editing activity was quantified based on ^{32}P-labeled U spots migrating slower than ^{32}P–U spots. Editing activity is influenced by the concentrations of magnesium and potassium. The addition of ATP strikingly enhanced the editing activity. The $[\alpha^{32}P]$-labeled C-to-U change indicated that the α-phosphate group of the edited residue was retained during the editing reaction; evidently, the C-to-U conversion in chloroplasts occurs by cytidine deamination, and not by nucleotide substitution, as suggested for plant mitochondria (Rajasekhar and Mulligan 1993). After the editing reaction, cDNA was amplified from the isolated mRNA substrate, and its sequence analysis revealed that C-to-U conversion occurred exclusively at the authentic editing site, confirming that the in vitro system supports accurate RNA editing of the *psbL* mRNA.

As described above, *cis*-acting elements for editing were identified in vivo for tobacco *psbL* and *ndhB* mRNAs (sites IV and V; Chaudhuri and Maliga 1996; Bock

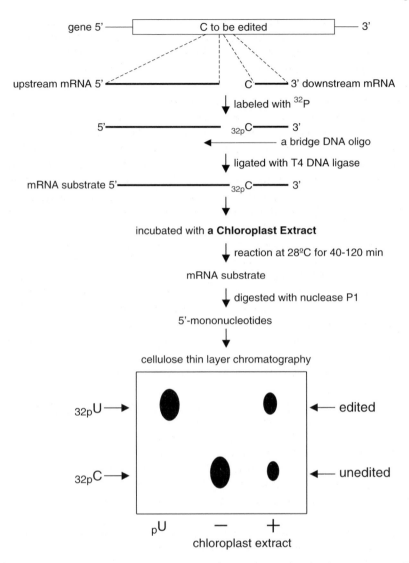

Fig. 1 In vitro assay to detect RNA editing activity using an mRNA substrate labeled with ^{32}P at the C residue to be edited

et al. 1996). In order to confirm whether the in vitro editing system depends on these *cis*-elements, mutational analysis of tobacco *psbL* and *ndhB* mRNAs was performed. Substitution of either the 16-nt upstream sequence or the 9-nt downstream sequence by a vector sequence abolished editing, indicating that both upstream and downstream regions are essential for editing of *psbL* mRNAs. A similar analysis of the *ndhB* mRNA showed that replacement of only the upstream, but not the downstream, region inhibited editing completely. These results confirmed those obtained from the in vivo analysis using transplastomic tobacco plants. To examine the involvement of *trans*-acting factors in editing, competition analyses using 25-nt oligoribonucleotides corresponding to upstream and downstream regions of the editing sites in *psbL* or *ndhB* mRNAs were carried out. The editing of *psbL* and *ndhB* mRNAs was arrested by the addition of excess amounts of the upstream competitors, but not by the downstream competitors. These results strongly suggest the existence of a *trans*-acting factor(s) specifically interacting with the upstream region of each mRNA. The competitor RNA was then exchanged. The editing of *psbL* mRNA was not arrested by an excess of the competitor for *ndhB* mRNA, and that of *ndhB* was not inhibited by the *psbL* competitor, suggesting that the *trans*-factors are site-specific. These results again completely confirmed those obtained from the in vivo analysis (Hirose and Sugiura 2001).

The most intriguing question about *trans*-factors for RNA editing in chloroplasts is whether they contain an RNA component(s). Editing activity did not decrease after treatment of the chloroplast extract with micrococcal nuclease. Furthermore, attempts to detect RNA molecules interacting with *psbL* mRNAs by cross-linking in the presence of AMT (4′-aminomethyl-4,5′, 8-trimethyl-psoralen), which forms covalent adducts after irradiation with long-wavelength UV light, yielded negative results. Although these experiments are indirect and preliminary, the *trans*-factor is likely to be a protein, rather than RNA. In vivo analyses could not provide clear conclusions about the involvement of RNA components (Bock and Maliga 1995; Hegeman et al. 2005b). To detect proteins interacting with the upstream *cis*-element, UV cross-linking experiments were carried out using the *psbL* mRNA substrate labeled at the −6 residue relative to the C to be edited (not at the editing C residue, as above). At least five chloroplast proteins ranging in size from 25 to 45 kDa were detected. Competition analysis revealed that the 25-kDa protein (p25) binds specifically to *psbL* mRNA, its binding was arrested by an excess amount of the upstream *psbL* competitor RNA, but not by the other RNAs, and the characteristics of p25 binding to *psbL* mRNA correspond to that of editing activity. Similar experiments showed that p25 did not bind to the *ndhB* mRNA, indicating that p25 is a *trans*-factor specific for the editing of *psbL* mRNA (Hirose and Sugiura 2001).

Based on their sizes, the additional cross-linked proteins of 28–33 kDa were considered to be chloroplast RNA-binding proteins previously isolated from tobacco chloroplasts (cp28, cp29A, cp29B, cp31, and cp33; Li and Sugiura 1990; Ye et al. 1991). Each is an abundant stromal protein that possesses two RNA recognition motifs (RRM), or consensus-type RNA-binding domains (CS-RBD), and an N-terminal acidic domain, and is associated with various chloroplast RNA species including mRNAs. Therefore, the effect of antibodies against each cp protein on the in vitro editing reactions was examined. The editing of both *psbL* and *ndhB* mRNAs was inhibited only by the addition of anti-cp31. Immunodepletion of cp31 from the chloroplast extract resulted in the inhibition of *psbL* mRNA editing, and the addition of recombinant cp31 into the cp31-depleted extract restored editing activity. These results indicated that cp31 is an additional essential factor involved in the editing of *psbL* and *ndhB* mRNAs, suggesting that cp31 is a common factor for at least a group of editing sites. This is the first protein identified to bind to *cis*-elements for editing, and its structure has also been elucidated. The mature cp31 is 244 amino acids long, and consists of the acidic N-terminal domain (64 amino acids), two RRMs (83 amino acids each) separated by 11 amino acids, and the C-terminal three amino acids (Li and Sugiura 1990). These two RRMs are highly similar to those of a large number of proteins that are involved in RNA processing and RNA metabolism. To investigate the function of the N-terminal domain of cp31, recombinant cp31 lacking this domain was prepared. When the deleted cp31 was added to the cp31-depleted extract, editing was hardly detected, indicating this domain is necessary for the function of cp31 in editing (Hirose and Sugiura 2001). The C-to-U editing of mammalian apolipoprotein-B mRNA is catalyzed by an enzyme complex that recognizes an 11-nt mooring sequence downstream of the C to be edited. The complex contains apobec-1 (a cytidine deaminase) and apobec-1 complementation factor (ACF), the RNA-binding subunit that binds to the mooring sequence and interacts with apobec-1 (Mehta and Driscoll 2002). ACF consists of three RRMs and N- and C-terminal domains, and both terminal domains are required for complementing activity and high-affinity binding to the mRNA. Similarities in structure and its characteristics in vitro with ACF suggest that cp31 complements a site-specific factor to edit a group of mRNAs in chloroplasts.

The original procedure for preparing chloroplast extracts was then modified, and the improved methods provided much more active extracts not only from tobacco, but also from pea (Miyamoto et al. 2002; Hirose et al. 2004; Nakajima and Mulligan 2005). This procedure was successfully applied for preparing plant mitochondrial extracts (Takenaka and Brennicke 2003), and for *Arabidopsis* and maize chloroplast extracts (Hegeman et al. 2005a; Hayes et al. 2006). Using the improved extract with site-specifically labeled mRNA substrates, editing of tobacco *psbE* and *petB* mRNA was then analyzed in vitro, and *cis*-elements were defined for the 10-nt sequence from −15 to −6 for *psbE* mRNA, and the 15-nt sequence from −20 to −6 for *petB* mRNA (Miyamoto et al. 2002). UV cross-linking assays with mRNA substrates specifically labeled at the center of the *cis*-elements indicated again that distinct proteins bind specifically to each *cis*-element, a 56-kDa protein to the *psbE* site, and a 70-kDa species to the *petB* site. Pea chloroplasts lack the corresponding editing site in *psbE* (T is already present in the DNA). Parallel in vitro analyses with tobacco and pea extracts revealed that the pea plant has no editing activity for *psbE* mRNAs and lacks the 56-kDa protein, whereas *petB* mRNAs are edited and the 70-kDa protein is also present (Fig. 2). Therefore, co-evolution of an editing site and its cognate *trans*-factor was demonstrated biochemically in *psbE* mRNA editing between tobacco and pea plants.

Further in vitro editing assays with mutated mRNA substrates clearly indicated that the *cis*-elements (*trans*-factor binding sites) described above are not sufficient, and that the 5-nt region immediately upstream from the editing sites is also essential for editing. Point mutation analysis showed that a specific sequence (not as a distance) is required in the 5-nt region. Using *psbE* and *petB* mRNAs labeled either at the center of the *cis*-element or at the editing site, the site-specific factors were

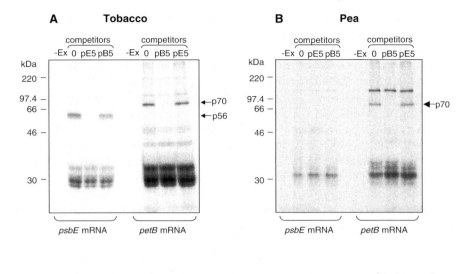

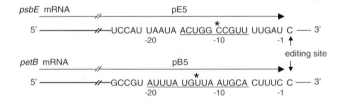

Fig. 2 Identification of site-specific protein factors for RNA editing. **A** Detection of tobacco chloroplast proteins (p56 and p70) bound to the *cis*-elements of *psbE* and *petB* mRNAs by UV cross-linking. **B** Detection of pea chloroplast proteins (p70) bound as in **A**. No p56 band is seen, at no corresponding site in pea. **C** *Long arrows* indicate unlabeled competitor RNAs for *psbE* mRNA (pE5) and for *petB* mRNA (pB5) that correspond to sequences 5′ at the editing sites. *Underlines* indicate *cis*-elements and *asterisks* ^{32}P-labeled residues. The mRNA was incubated for 1 h at 28 °C in the editing reaction mixture, and then UV irradiated. The mixture was treated with RNase A, followed by PAGE. A hundredfold excess of the competitors was added. *−Ex* Without chloroplast extracts, *0* without competitor. Protein size markers are shown at the *left*. Modified from Figs. 5 and 6 of Miyamoto et al. (2002), with permission

found to be cross-linked with nucleotides at both positions. Mutations of nucleotides in the proximal region of the editing site revealed a correlation between editing activity and cross-linking efficiency of factors with the editing site, even though cross-linking with the upstream *cis*-element was unaffected. These observations suggest that the site-specific factor binds stably to the upstream *cis*-element, whereas it interacts weakly with the editing site. This finding raises the possibility that the site-specific factor is involved in both site-determination and C-to-U conversion in at least some chloroplast RNA editing (Miyamoto et al. 2004).

The above in vitro system required mRNA substrates specifically labeled with ^{32}P at editing sites (or other sites of interest), which is technically difficult and expensive to prepare. Therefore, a simple poisoned primer extension assay was devised to replace the use of radioactive substrates, and active chloroplast extracts were obtained from *Arabidopsis* (Hegeman et al. 2005a). Using this new system, editing of *Arabidopsis psbE* mRNA was found to require ATP, CTP, or dCTP, suggesting the involvement of RNA helicase activity, and to be inhibited by a zinc-specific chelator, a sensitivity typical of zinc-dependent cytidine deaminases. Similarly, the sequence GCCGUU, which occurs 5' of the editing site, was discovered to be critical for editing of tobacco *psbE* mRNAs (Hayes and Hanson 2007). An alternative method was also developed using primer extension assays with fluorescent-labeled ddATP and ddGTP (Sasaki et al. 2006). Primers are designed so that the 3' ends correspond to 1 nt after the editing site (position +2). ddATP is incorporated when mRNAs are edited, while ddGTP is added when unedited, and fluorescence is detected by an automated DNA sequencer. Editing of all the 17 sites found in tobacco *ndhA* to *G* mRNAs was analyzed by the new assay; editing efficiencies were found to vary from null to 35% (the *psbE* editing efficiency was 86%). Among these, two efficiently edited sites, the *ndhB* site 2 and *ndhF* site 2, were subjected to scanning mutagenesis; the *cis*-element of the *ndhB* site 2 was found to be from −10 to −6. Unlike the other sites so far examined, the *ndhF* site 2 apparently possesses two separated *cis*-elements, i.e., sequences from −40 to −36, and from −15 to −6; the situation is similar to that found in the *atp9* (site 1) mRNA editing in pea mitochondria (Takenaka et al. 2004).

6   Summary and Perspectives

Chloroplast pre-mRNAs from angiosperm plants undergo C-to-U editing at approximately 30 specific sites, but no consensus motifs, such as in the case of splice sites, have been identified. This makes the prediction of editing sites difficult. Searching for potential editing sites based on amino acid sequence alignment was applicable only for protein-coding regions, and was inefficient: only 19 true edited sites (4.2%) of 451 predicted sites in pea chloroplasts have been identified (Inada et al. 2004), and editing in the identified sites logically restored codons for evolutionarily conserved amino acids. Therefore, it is not reasonable to assume that editing is always essential for protein function. Some editing events were proven to be functionally important, while some others were not (mostly identified by chance), and the majority of editing events remained unanalyzed for their functional significance. The list of identified editing sites is not complete, because searching has not accounted for editing sites outside the protein-coding regions, codons for non-conserved amino acids, and tRNA, rRNA and other non-coding RNA transcripts. Efforts have to be continued to identify additional editing sites in order to understand the full genetic information in the chloroplast genome.

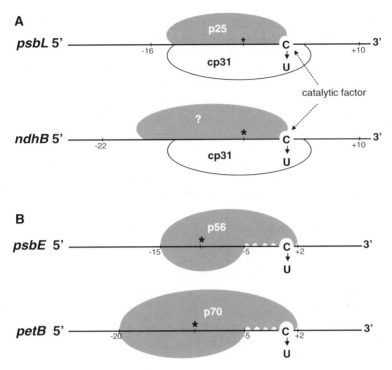

Fig. 3 Models for the mechanism of RNA editing in tobacco chloroplasts. **A** A site-specific *trans*-acting factor (p56 or p70) binds a *cis*-acting element (the 10 nt from −15 to −6 for *psbE*, and the 15 nt from −20 to −6 for *petB*) in a sequence-specific manner, and the factor then interacts with an editing site also in a sequence-specific manner. This interaction is most likely necessary for C-to-U conversion, possibly by allowing contact of the putative catalytic domain of the site-specific factor with the target C residue. **B** A site-specific *trans*-acting factor (p25 for *psbL* mRNA) binds to the upstream *cis*-acting element of an editing site. One of the abundant chloroplast RNA-binding proteins (cp31), probably a common factor, also binds close to every editing site. Complexes including these proteins may recruit the catalytic factor of C-to-U conversion to the editing sites. *Numbers* represent *cis*-element positions defined by transplastomic experiments (Chaudhuri and Maliga 1996; Bock et al. 1996). Redrawn based on the models in Hirose and Sugiura (2001) and in Miyamoto et al. (2004)

The mechanism of RNA editing has been studied by in vivo and in vitro techniques. Using transgenic technologies with tobacco chloroplasts, *cis*-elements for editing were found, and the existence of cognate *trans*-acting factors was suggested. A genetic approach with *Arabidopsis* has the potential to identify nuclear genes involved in chloroplast RNA editing. However, the in vivo method is laborious and time-consuming, and is not suitable to elucidate the biochemical mechanism of editing reactions. In vitro RNA editing systems have enabled us to analyze *cis*-elements in more detail, and to identify *trans*-factors that bind directly to the *cis*-elements, as well as auxiliary components and cofactors, such as ATP. A drawback to any in vitro system is the reproducibility of active extract preparations.

In vitro artifacts are always a major issue, and one should eliminate these by parallel experiments designed appropriately and thoroughly.

Based on available data, it is strongly suggested that RNA editing in chloroplasts proceeds by multiple mechanisms. Figure 3 shows two models, a multiple subunit complex including a site-recognition factor, cytidine deaminase and other accessory factors (Fig. 3A), and a single-component or its dimeric or oligomeric form (Fig. 3B). Currently, the most puzzling question is the origin and evolution of RNA editing in the chloroplast genetic system, as some of the editing events seem nonessential for proper expression of chloroplast genes. Comparative phylogenetic analysis showed that neither editing frequencies nor editing patterns correlate with the phylogenetic tree. Elucidation of the molecular mechanism of editing will help our understanding of the origin and evolution of this process.

Acknowledgements I thank all the members in my laboratory who have been working on RNA editing in chloroplasts. I especially thank Tadamasa Sasaki for helping to compile editing sites, and Yoko Shimomura for assistance in preparing the text and figures. I am most grateful to the Sugiyama Human Research Center, Sugiyama Jogakuen University (Nagoya, Japan) for providing me with a pleasant environment to write this manuscript. Research on RNA editing in my laboratory is supported by a Grant-in-Aid from the Japan Society for Promotion of Sciences.

References

Bock R (2000) Sense from nonsense: how the genetic information of chloroplasts is altered by RNA editing. Biochimie 82:549–557

Bock R (2001) RNA editing in plant mitochondria and chloroplasts. In: Bass BL (ed) RNA editing. Oxford University Press, New York, pp 38–60

Bock R (2004) Studying RNA editing in transgenic chloroplasts of higher plants. In: Gott JM (ed) Methods in Molecular Biology vol 265. Humana Press, Totowa, pp 345–356

Bock R, Koop HU (1997) Extraplastidic site-specific factors mediate RNA editing in chloroplasts. EMBO J 16:3282–3288

Bock R, Maliga P (1995) In vivo testing of a tobacco plastid DNA segment for guide RNA function in psbL editing. Mol Gen Genet 247:439–443

Bock R, Kossel H, Maliga P (1994) Introduction of a heterologous editing site into the tobacco plastid genome: the lack of RNA editing leads to a mutant phenotype. EMBO J 13:4623–4628

Bock R, Hermann M, Kossel H (1996) In vivo dissection of cis-acting determinants for plastid RNA editing. EMBO J 15:5052–5059

Bock R, Albertazzi F, Freyer R, Fuchs M, Ruf S, Zeltz P, Maier RM (1997a) Transcript editing in chloroplasts of higher plants. In: Schenk HEA, Herrmann RG, Jeon KW, Müller NE, Schwemmler W (eds) Eukaryotism and symbiosis. Springer, Berlin Heidelberg New York, pp 123–137

Bock R, Hermann M, Fuchs M (1997b) Identification of critical nucleotide positions for plastid RNA editing site recognition. RNA 3:1194–1200

Calsa T Jr, Carraro DM, Benatti MR, Barbosa AC, Kitajima JP, Carrer H (2004) Structural features and transcript-editing analysis of sugarcane (*Saccharum officinarum* L.) chloroplast genome. Curr Genet 46:366–373

Chateigner-Boutin AL, Hanson MR (2002) Cross-competition in transgenic chloroplasts expressing single editing sites reveals shared cis elements. Mol Cell Biol 22:8448–8456

Chateigner-Boutin AL, Hanson MR (2003) Developmental co-variation of RNA editing extent of plastid editing sites exhibiting similar cis-elements. Nucleic Acids Res 31:2586–2594

Chaudhuri S, Maliga P (1996) Sequences directing C to U editing of the plastid psbL mRNA are located within a 22 nucleotide segment spanning the editing site. EMBO J 15:5958–5964

Chaudhuri S, Carrer H, Maliga P (1995) Site-specific factor involved in the editing of the psbL mRNA in tobacco plastids. EMBO J 14:2951–2957

Corneille S, Lutz K, Maliga P (2000) Conservation of RNA editing between rice and maize plastids: are most editing events dispensable? Mol Gen Genet 264:419–424

Del Campo EM, Sabater B, Martin M (2000) Transcripts of the ndhH-D operon of barley plastids: possible role of unedited site III in splicing of the ndhA intron. Nucleic Acids Res 28:1092–1098

Del Campo EM, Sabater B, Martin M (2002) Post-transcriptional control of chloroplast gene expression. Accumulation of stable psaC mRNA is due to downstream RNA cleavages in the ndhD gene. J Biol Chem 277:36457–36464

Duff RJ, Moore FB (2005) Pervasive RNA editing among hornwort rbcL transcripts except *Leiosporoceros*. J Mol Evol 61:571–578

Fiebig A, Stegemann S, Bock R (2004) Rapid evolution of RNA editing sites in a small non-essential plastid gene. Nucleic Acids Res 32:3615–3622

Freyer R, Kiefer-Meyer MC, Kossel H (1997) Occurrence of plastid RNA editing in all major lineages of land plants. Proc Natl Acad Sci USA 94:6285–6290

Giege P, Brennicke A (1999) RNA editing in *Arabidopsis* mitochondria effects 441 C to U changes in ORFs. Proc Natl Acad Sci USA 96:15324–15329

Halter CP, Peeters NM, Hanson MR (2004) RNA editing in ribosome-less plastids of iojap maize. Curr Genet 45:331–337

Hayes ML, Hanson MR (2007) Identification of a sequence motif critical for editing of a tobacco chloroplast transcript. RNA 13:281–288

Hayes ML, Reed ML, Hegeman CE, Hanson MR (2006) Sequence elements critical for efficient RNA editing of a tobacco chloroplast transcript in vivo and in vitro. Nucleic Acids Res 34:3742–3754

Hegeman CE, Hayes ML, Hanson MR (2005a) Substrate and cofactor requirements for RNA editing of chloroplast transcripts in *Arabidopsis* in vitro. Plant J 42:124–132

Hegeman CE, Halter CP, Owens TG, Hanson MR (2005b) Expression of complementary RNA from chloroplast transgenes affects editing efficiency of transgene and endogenous chloroplast transcripts. Nucleic Acids Res 33:1454–1464

Hermann M, Bock R (1999) Transfer of plastid RNA-editing activity to novel sites suggests a critical role for spacing in editing-site recognition. Proc Natl Acad Sci USA 96:4856–4861

Hirose T, Sugiura M (1997) Both RNA editing and RNA cleavage are required for translation of tobacco chloroplast ndhD mRNA: a possible regulatory mechanism for the expression of a chloroplast operon consisting of functionally unrelated genes. EMBO J 16:6804–6811

Hirose T, Sugiura M (2001) Involvement of a site-specific trans-acting factor and a common RNA-binding protein in the editing of chloroplast mRNAs: development of a chloroplast in vitro RNA editing system. EMBO J 20:1144–1152

Hirose T, Fan H, Suzuki JY, Wakasugi T, Tsudzuki T, Kossel H, Sugiura M (1996) Occurrence of silent RNA editing in chloroplasts: its species specificity and the influence of environmental and developmental conditions. Plant Mol Biol 30:667–672

Hirose T, Kusumegi T, Tsudzuki T, Sugiura M (1999) RNA editing sites in tobacco chloroplast transcripts: editing as a possible regulator of chloroplast RNA polymerase activity. Mol Gen Genet 262:462–467

Hirose T, Miyamoto T, Obokata J, Sugiura M (2004) In vitro RNA editing systems from higher plant chloroplasts. In: Gott JM (ed) Methods in Molecular Biology, vol 265. Humana Press, Totowa, pp 333–344

Hoch B, Maier RM, Appel K, Igloi GL, Kossel H (1991) Editing of a chloroplast mRNA by creation of an initiation codon. Nature 353:178–180

Inada M, Sasaki T, Yukawa M, Tsudzuki T, Sugiura M (2004) A systematic search for RNA editing sites in pea chloroplasts: an editing event causes diversification from the evolutionarily conserved amino acid sequence. Plant Cell Physiol 45:1615–1622

Kahlau S, Aspinall S, Gray JC, Bock R (2006) Sequence of the tomato chloroplast DNA and evolutionary comparison of solanaceous plastid genomes. J Mol Evol 63:194–207

Karcher D, Bock R (1998) Site-selective inhibition of plastid RNA editing by heat shock and antibiotics: a role for plastid translation in RNA editing. Nucleic Acids Res 26:1185–1190

Kotera E, Tasaka M, Shikanai T (2005) A pentatricopeptide repeat protein is essential for RNA editing in chloroplasts. Nature 433:326–330

Kudla J, Bock R (1999) RNA editing in an untranslated region of the Ginkgo chloroplast genome. Gene 234:81–86

Kugita M, Yamamoto Y, Fujikawa T, Matsumoto T, Yoshinaga K (2003) RNA editing in hornwort chloroplasts makes more than half the genes functional. Nucleic Acids Res 31:2417–2423

Li YQ, Sugiura M (1990) Three distinct ribonucleoproteins from tobacco chloroplasts: each contains a unique amino terminal acidic domain and two ribonucleoprotein consensus motifs. EMBO J 9:3059–3066

Lurin C, Andres C, Aubourg S, Bellaoui M, Bitton F, Bruyere C, Caboche M, Debast C, Gualberto J, Hoffmann B, Lecharny A, Le Ret M, Martin-Magniette ML, Mireau H, Peeters N, Renou JP, Szurek B, Taconnat L, Small I (2004) Genome-wide analysis of *Arabidopsis* pentatricopeptide repeat proteins reveals their essential role in organelle biogenesis. Plant Cell 16:2089–2103

Lutz KA, Maliga P (2001) Lack of conservation of editing sites in mRNAs that encode subunits of the NAD(P)H dehydrogenase complex in plastids and mitochondria of *Arabidopsis thaliana*. Curr Genet 40:214–219

Lutz KA, Bosacchi MH, Maliga P (2006) Plastid marker-gene excision by transiently expressed CRE recombinase. Plant J 45:447–456

Maier RM, Hoch B, Zeltz P, Kossel H (1992) Internal editing of the maize chloroplast ndhA transcript restores codons for conserved amino acids. Plant Cell 4:609–616

Maier RM, Neckermann K, Igloi GL, Kossel H (1995) Complete sequence of the maize chloroplast genome: gene content, hotspots of divergence and fine tuning of genetic information by transcript editing. J Mol Biol 251:614–628

Mehta A, Driscoll DM (2002) Identification of domains in apobec-1 complementation factor required for RNA binding and apolipoprotein-B mRNA editing. RNA 8:69–82

Miyamoto T, Obokata J, Sugiura M (2002) Recognition of RNA editing sites is directed by unique proteins in chloroplasts: biochemical identification of cis-acting elements and trans-acting factors involved in RNA editing in tobacco and pea chloroplasts. Mol Cell Biol 22:6726–6734

Miyamoto T, Obokata J, Sugiura M (2004) A site-specific factor interacts directly with its cognate RNA editing site in chloroplast transcripts. Proc Natl Acad Sci USA 101:48–52

Mulligan RM (2004) RNA editing in plant organelles. In: Daniell H, Chase CD (eds) Molecular biology and biotechnology of plant organelles. Kluwer, Dordrecht, pp 239–260

Nakajima Y, Mulligan RM (2005) Nucleotide specificity of the RNA editing reaction in pea chloroplasts. J Plant Physiol 162:1347–1354

Okuda K, Nakamura T, Sugita M, Shimizu T, Shikanai T (2006) A pentatricopeptide repeat protein is a site recognition factor in chloroplast RNA editing. J Biol Chem 281:37661–37667

Okuda K, Myouga F, Motohashi R, Shinozaki K, Shikanai T (2007) Conserved domain structure of pentatricopeptide repeat proteins involved in chloroplast RNA editing. Proc Natl Acad Sci USA 104:8178–8183

Peeters NM, Hanson MR (2002) Transcript abundance supercedes editing efficiency as a factor in developmental variation of chloroplast gene expression. RNA 8:497–511

Rajasekhar VK, Mulligan RM (1993) RNA editing in plant mitochondria: [alpha]-phosphate is retained during C-to-U conversion in mRNAs. Plant Cell 5:1843–1852

Reed ML, Hanson MR (1997) A heterologous maize rpoB editing site is recognized by transgenic tobacco chloroplasts. Mol Cell Biol 17:6948–6952

Reed ML, Lyi SM, Hanson MR (2001a) Edited transcripts compete with unedited mRNAs for trans-acting editing factors in higher plant chloroplasts. Gene 272:165–171

Reed ML, Peeters NM, Hanson MR (2001b) A single alteration 20 nt 5′ to an editing target inhibits chloroplast RNA editing in vivo. Nucleic Acids Res 29:1507–1513

Sasaki T (2005) A systematic identification of RNA editing sites in *Nicotiana* chloroplasts and analysis of *cis*-elements for editing using a new *in vitro* system. PhD Thesis, Nagoya City University, Japan

Sasaki Y, Kozaki A, Ohmori A, Iguchi H, Nagano Y (2001) Chloroplast RNA editing required for functional acetyl-CoA carboxylase in plants. J Biol Chem 276:3937–3940

Sasaki T, Yukawa Y, Miyamoto T, Obokata J, Sugiura M (2003) Identification of RNA editing sites in chloroplast transcripts from the maternal and paternal progenitors of tobacco (*Nicotiana tabacum*): comparative analysis shows the involvement of distinct trans-factors for ndhB editing. Mol Biol Evol 20:1028–1035

Sasaki T, Yukawa Y, Wakasugi T, Yamada K, Sugiura M (2006) A simple *in vitro* RNA editing assay for chloroplast transcripts using fluorescent dideoxynucleotides: distinct types of sequence elements required for editing of *ndh* transcripts. Plant J 47:802–810

Schmitz-Linneweber C, Tillich M, Herrmann RG, Maier RM (2001) Heterologous, splicing-dependent RNA editing in chloroplasts: allotetraploidy provides trans-factors. EMBO J 20:4874–4883

Schmitz-Linneweber C, Regel R, Du TG, Hupfer H, Herrmann RG, Maier RM (2002) The plastid chromosome of *Atropa belladonna* and its comparison with that of *Nicotiana tabacum*: the role of RNA editing in generating divergence in the process of plant speciation. Mol Biol Evol 19:1602–1612

Schmitz-Linneweber C, Kushnir S, Babiychuk E, Poltnigg P, Herrmann RG, Maier RM (2005) Pigment deficiency in nightshade/tobacco cybrids is caused by the failure to edit the plastid ATPase alpha-subunit mRNA. Plant Cell 17:1815–1828

Shikanai T (2006) RNA editing in plant organelles: machinery, physiological function and evolution. Cell Mol Life Sci 63:698–708

Sugiura M (1992) The chloroplast genome. Plant Mol Biol 19:149–168

Svab Z, Maliga P (1993) High-frequency plastid transformation in tobacco by selection for a chimeric aadA gene. Proc Natl Acad Sci USA 90:913–917

Takenaka M, Brennicke A (2003) In vitro RNA editing in pea mitochondria requires NTP or dNTP, suggesting involvement of an RNA helicase. J Biol Chem 278:47526–47533

Takenaka M, Neuwirt J, Brennicke A (2004) Complex cis-elements determine an RNA editing site in pea mitochondria. Nucleic Acids Res 32:4137–4144

Tillich M, Funk HT, Schmitz-Linneweber C, Poltnigg P, Sabater B, Martin M, Maier RM (2005) Editing of plastid RNA in *Arabidopsis thaliana* ecotypes. Plant J 43:708–715

Tillich M, Poltnigg P, Kushnir S, Schmitz-Linneweber C (2006a) Maintenance of plastid RNA editing activities independently of their target sites. EMBO Rep 7:308–313

Tillich M, Lehwark P, Morton BR, Maier UG (2006b) The evolution of chloroplast RNA editing. Mol Biol Evol 23:1912–1921

Tsudzuki T, Wakasugi T, Sugiura M (2001) Comparative analysis of RNA editing sites in higher plant chloroplasts. J Mol Evol 53:327–332

Wakasugi T, Hirose T, Horihata M, Tsudzuki T, Kossel H, Sugiura M (1996) Creation of a novel protein-coding region at the RNA level in black pine chloroplasts: the pattern of RNA editing in the gymnosperm chloroplast is different from that in angiosperms. Proc Natl Acad Sci USA 93:8766–8770

Wakasugi T, Tsudzuki T, Sugiura M (2001) The genomics of land plant chloroplasts: gene content and alteration of genomic information by RNA editing. Photosynth Res 70:107–118

Wolf PG, Rowe CA, Hasebe M (2004) High levels of RNA editing in a vascular plant chloroplast genome: analysis of transcripts from the fern *Adiantum capillus-veneris*. Gene 339:89–97

Ye LH, Li YQ, Fukami-Kobayashi K, Go M, Konishi T, Watanabe A, Sugiura M (1991) Diversity of a ribonucleoprotein family in tobacco chloroplasts: two new chloroplast ribonucleoproteins and a phylogenetic tree of ten chloroplast RNA-binding domains. Nucleic Acids Res 19:6485–6490

Yoshinaga K, Iinuma H, Masuzawa T, Uedal K (1996) Extensive RNA editing of U to C in addition to C to U substitution in the rbcL transcripts of hornwort chloroplasts and the origin of RNA editing in green plants. Nucleic Acids Res 24:1008–1014

Zandueta-Criado A, Bock R (2004) Surprising features of plastid ndhD transcripts: addition of non-encoded nucleotides and polysome association of mRNAs with an unedited start codon. Nucleic Acids Res 32:542–550

Zeltz P, Hess WR, Neckermann K, Borner T, Kossel H (1993) Editing of the chloroplast rpoB transcript is independent of chloroplast translation and shows different patterns in barley and maize. EMBO J 12:4291–4296

Zeng WH, Liao SC, Chang CC (2007) Identification of RNA editing sites in chloroplast transcripts of *Phalaenopsis aphrodite* and comparative analysis with those of other seed plants. Plant Cell Physiol 48:362–368

Zito F, Kuras R, Choquet Y, Kossel H, Wollman FA (1997) Mutations of cytochrome b6 in *Chlamydomonas reinhardtii* disclose the functional significance for a proline to leucine conversion by petB editing in maize and tobacco. Plant Mol Biol 33:79–86

Working Together: the RNA Editing Machinery in *Trypanosoma brucei*

Jason Carnes and Kenneth Stuart (✉)

"Why shouldn't truth be stranger than fiction? Fiction, after all, has to make sense."

Mark Twain

Abstract The proteins in the ~20S editosome function in a coordinated fashion to catalyze the precise insertion and deletion of uridine nucleotides (Us) in mitochondrial pre-mRNAs, thereby generating functional mRNAs. Many

Seattle Biomedical Research Institute, 307 Westlake Ave N, Suite 500, Seattle, WA 98109, USA, and Department of Pathobiology, University of Washington, Seattle, WA 98195, USA; kstuart@u.washington.edu

H.U. Göringer (ed.), *RNA Editing. Nucleic Acids and Molecular Biology 20*
© Springer-Verlag Berlin Heidelberg 2008

catalytic components of the editosome have been identified, but the abundance of apparently non-catalytic components is a conspicuous reminder that the editosome is much more than the sum of its catalytic activities. As the functions of the various editosome proteins are elucidated, the importance of how they interact in vivo is becoming more evident. Indeed, ~20S editosomes are compositionally heterogeneous in vivo, and higher-order interactions with proteins of other complexes affect editing. Here, we review the general process of editing, and discuss how the known editosome proteins interact and function to produce translatable mRNAs in vivo.

1 Introduction

1.1 RNA Editing and Kinetoplastids

RNA editing is characteristic of the order Kinetoplastida, reflecting the unusual mitochondrial DNA in these unicellular protozoans. This DNA was seen as an intensely staining body that early cytologists named the kinetoplast for its proximity to the motile flagellum. This mitochondrial DNA, which is also called kinetoplast DNA (kDNA), exists as a complex network that is composed of two types of DNA: maxicircle DNA and minicircle DNA. Maxicircle DNA, like most other mitochondrial DNAs, encodes mitochondrial rRNAs and proteins of the oxidative phosphorylation system. However, most of the pre-mRNAs require post-transcriptional editing to become translatable mRNAs. Minicircles encode guide RNAs (gRNAs), which specify the final edited sequence as described in Sections 1.2 and 1.3. Most kinetoplastids have tens of identical maxicircles that are intercatenated with thousands of minicircles into a single, basket-like structure that lies within the single mitochondrion at a position across from the flagellar basal body.

Kinetoplastids are a diverse group of flagellated protozoans that diverged very early in the eukaryotic lineage. The order includes *Trypanosoma* and *Leishmania* species, numerous other pathogens, and some non-pathogenic species. While all of these organisms edit their mitochondrial mRNAs, the maxicircle and minicircle sequences have diverged, resulting in some differences in which mRNAs get edited, the extent of their editing, and the sequence of the final edited mRNAs. Editing is most extensive in *Trypanosoma* species, which are the main focus of this chapter.

1.2 Editing Description

RNA editing in trypanosomes remodels pre-mRNAs by the insertion and deletion of uridine nucleotides (Us) through a series of coordinated catalytic steps. These steps are performed by a multi-protein complex that sediments at ~20S in glycerol

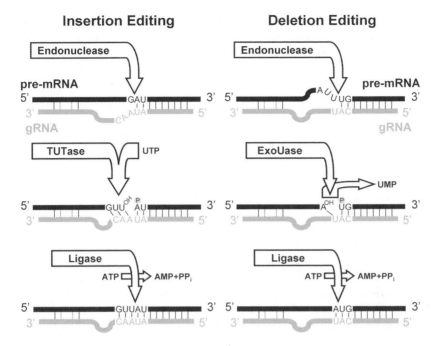

Fig. 1 General mechanism of catalytic events in U insertion and U deletion RNA editing. Interactions between guide RNAs (gRNAs; *grey strands*) and pre-mRNAs (*black strands*) include both Watson-Crick base-pairs (*solid lines*) and G:U base-pairs (*dotted lines*), and determine the cleavage site and subsequent U insertion (*left panel*) or U deletion (*right panel*). The first catalytic step in editing is the endonucleolytic cleavage of the pre-mRNA immediately upstream of the anchor duplex, which consists of 6 to 10 base-pairs between pre-mRNA and gRNA. Using gRNA as a template, Us are added to the 3′ end of the 5′ mRNA fragment by TUTase at insertion editing sites, or Us are removed by exoUase at deletion editing sites. RNA ligase then ligates the 5′ and 3′ mRNA fragments to complete a single editing cycle. Multiple editing cycles are usually encoded by each gRNA template, and multiple gRNA templates can be required to completely edit an entire pre-mRNA. Frequently, the editing guided by one gRNA generates sequence complementary to the anchor region of the next gRNA, so that the progression of editing through a particular pre-mRNA proceeds 3′ to 5′ and cycles through multiple gRNAs

gradients, aptly called the ~20S editosome. Complexes with editing components also sediment at ~40S (see Sect. 4.4), but the ~20S editosome catalyzes the four basic steps of editing in vitro and mimics a single round of insertion or deletion editing (Kable et al. 1996; Seiwert et al. 1996), as outlined in Fig. 1. The first catalytic step is the endonucleolytic cleavage of a pre-mRNA. The cleavage site is 5′ to the short (~6–10 bp) "anchor" duplex that forms between the cognate pre-mRNA and guide RNA (gRNA) that is adjacent to the region to be edited. Following endonucleolytic cleavage, the gRNA sequence specifies U removal by a U-specific 3′ exonuclease (exoUase), or U addition by a 3′ terminal uridylyl transferase (TUTase). The two mRNA fragments are rejoined by RNA ligases, thus completing a single cycle of catalysis. Single rounds of insertion and deletion editing can be reproduced in vitro. However, editing is more complex in vivo, since each gRNA specifies multiple rounds of editing, and some gRNAs specify both insertion and deletion editing sites. Indeed, most mRNAs require multiple and often numerous gRNAs to be completely edited to the final functional mRNA sequence.

1.3   Substrate RNAs

Most mitochondrial pre-mRNAs require editing to become functional mRNAs. Some mRNAs require little editing, such as cytochrome oxidase subunit II (COII) mRNA that has a total of four Us inserted at three sites. This editing eliminates a frameshift, thus specifying a larger protein than that specified by the pre-edited RNA. Editing of other mRNAs is more extensive and creates initiation codons, as in apocytochrome b (CYb), into which a total of 34 Us are inserted at 13 sites. The created AUG initiation codon is upstream of an AUG codon that exists in *T. brucei*, although it may not be functional, since it does not occur in other species such as *Leishmania tarentolae* and *Crithidia fasciculata*. Yet other mRNAs are extensively edited, such as cytochrome oxidase subunit III (COIII). Fully edited COIII mRNA has 547 Us inserted and 41 Us deleted, resulting in the creation of initiation and termination codons as well as the functional open reading frame. Proteins that are predicted from the edited mRNA sequences have been demonstrated in cellular material, thus showing that the mRNAs are functional even though they specify proteins with substantial phenylalanine content (Horváth et al. 2000). High phenylanine content is not surprising for proteins of the inner mitochondrial membrane, and is not peculiar to these proteins, since *Escherichia coli* has phenylalanine-rich proteins.

Pre-mRNAs are transcribed from maxicircle DNA as polycistronic primary transcripts. These transcripts are processed by cleavage and polyadenylation in addition to RNA editing. Cleavage of polycistronic pre-mRNAs is not well characterized, although it may be sequence-specific (Shu and Stuart 1994). The gRNAs that specify edited sequences are relatively small (~60 nt) and are encoded in the numerous minicircles. Each gRNA has a post-transcriptionally added 3' oligo(U) tail, a 5' region that forms the "anchor" duplex with pre-mRNA 3' to the region that will be edited, and a region between these that serves as a template directing the editing of several sites. Editing progresses 3' to 5' relative to the mRNA, and each gRNA specifies a "block" of sequence to be edited. Almost all mRNAs require the utilization of multiple gRNAs. Some gRNAs recognize the most 3' sequence of the unedited pre-mRNAs, and duplex with this sequence to begin the editing process. However, most gRNAs require editing of the pre-mRNA to create the sequence that forms an anchor duplex with its cognate gRNA. Thus, gRNAs are progressively utilized in an order that is determined by the ability of each to form an anchor duplex with the complementary sequence of the partially edited pre-mRNA. As a consequence, the final and functional mRNA is the product of multiple genes: those encoding the pre-mRNA as well as multiple gRNAs. An exception is COII, where the guiding sequence resides in the 3' UTR of the pre-mRNA.

A total of 3,030 Us are inserted and 322 deleted by editing in *T. brucei*. However, the gRNA coding capacity is more than sufficient for this. The minicircles of *T. brucei* encode ~1,200 different gRNAs, while an order of magnitude fewer gRNAs are required (Stuart et al. 1997). Sequencing of several gRNAs has revealed considerable gRNA sequence redundancy in *T. brucei* (Corell et al. 1993). In addition, editing is able to precede the cleavage of polycistronic pre-mRNAs, as is evident from the characterization of polycistronic sequences that are edited (Read et al. 1992). The role of the gRNA oligo(U) tail is uncertain, but it is essential, since knockdown of expression of KRET1, which catalyzes the oligo(U) addition to gRNAs (see Sect. 5.1), inhibits editing (Aphasizhev et al. 2002). The U tail has been suggested to function in associating the 3′ region of the gRNA with the mRNA upstream of the region of editing (Blum and Simpson 1990; Leung and Koslowsky 1999, 2001). While base-pairing was originally suggested to mediate this association, it now seems more likely that oligo(U) binding proteins are responsible, especially since such base-pairing potential is limited when the editing of most sites specified by a gRNA is completed.

1.4   Partially Edited mRNAs

Cellular RNA contains an abundance of partially edited RNAs (Abraham et al. 1988; Decker and Sollner-Webb 1990; Sturm and Simpson 1990). These RNAs are edited in their 3′ regions to varying extents and unedited in their 5′ regions, consistent with the 3′ to 5′ direction of editing. While a small fraction of these partially edited RNAs transition directly from fully edited to unedited sequence, the large majority have an incompletely edited region at the junction of the edited and unedited sequence. The incompletely edited junction region does not match either the unedited or final edited sequence, and its size can typically be specified by a single gRNA. The significance of these partially edited mRNA with the junctions is uncertain. Their abundance suggests that they are not aberrant side products, but this cannot be excluded. Some may encode alternative, i.e., truncated, proteins. In addition they may reflect the possibility that editing does not invariably proceed precisely 3′ to 5′ (Koslowsky et al. 1991).

1.5   Function of Edited mRNAs

Edited RNAs encode components of the oxidative phosphorylation system, as well as three proteins of undetermined function. In *T. brucei*, edited RNAs specify four subunits of NADH dehydrogenase (or NADH ubiquinone oxidoreductase; respiratory complex I), the apocytochrome b (CYb) component of the cytochrome bc_1 complex (complex III), subunits II and III (COII and COIII) of the cytochrome c oxidase complex (complex IV), subunit 6 of the ATP synthase complex (complex V), and a protein with limited predicted homology to ribosomal protein S12 (RPS12). The three proteins with undetermined functions are named CR3, CR4, and MURF2; the sequence of these proteins provides limited evidence that they could code for components of respiratory complex I (Stuart et al. 1997). Editing is developmentally regulated in *T. brucei* in a fashion that parallels the respiratory system changes that occur during the life cycle. In the bloodstream stage of the life cycle, edited CYb and COII mRNAs are absent while those of complex I are abundant, reflecting the use of glycolysis for energy production. In the insect stage, where energy is generated by cytochrome-mediated oxidative phosphorylation, edited CYb and COII mRNAs are abundant while those for complex I are less so. The mechanism that mediates this regulation is unknown, but is not through gRNA abundance (Riley et al. 1995). Which RNAs are edited, and the extent of their editing, differ among the various kinetoplastids (Stuart et al. 1997). In addition, the regulation of editing in other species is less obvious and less studied than in *T. brucei*. Overall, editing appears to be involved in regulating the function of the oxidative phosphorylation system, which must be coordinated with other mechanisms that control energy generation and other life-cycle stage-specific processes, especially in *T. brucei*.

1.6   Components of the Editosome

The ~20S editosome was initially enriched by different biochemical methods, and later by immunopurification using monoclonal antibodies and by tandem affinity purification (TAP) tags. TAP-tags allow specific and efficient isolation of complexes, and portions thereof, via a tag that contains protein A and calmodulin binding domains separated by a TEV protease cleavage site (Rigaut et al. 1999). The complexes are purified following introduction of conditionally expressed

Table 1

[a] Editosome proteins are at the top, and other proteins implicated in RNA editing are at the bottom. Because numerous nomenclature systems exist in RNA editing literature, current names for each protein are followed by their former names. The function of many proteins has been determined experimentally, while others are based on sequence predictions (denoted by an asterisk). The function "Interaction" is ascribed to proteins that bind to RNA/protein and lack known catalytic activity. Sequence motifs that are found in each protein are listed in the far right column, and question marks next to a motif indicate low similarity scores. OB-fold, Oligonucleotide binding fold; zinc-finger, C2H2-type zinc-finger domain; RNase III, endoribonuclease motif from RNase III; dsRBM, double-stranded RNA binding motif; U1-like, U1-like zinc-finger motif; Pumilio domains, RNA binding motifs; ligase, signature ligase motif; tau and K, putative microtubule associated tau and kinesin light chain domains, respectively; 5′-3′ exo, 5′ to 3′ exoribonuclease motif; endo/exo/phos, endonuclease/exonuclease/phosphatase domain; NT, nucleotidyl transferase domain; PAP-core and PAP-assoc, poly(A) polymerase core and associated domains, respectively; RGG, arginine-glycine-glycine motif; Arg-rich, arginine-rich domain. See text for references

Table 1 Proteins involved in editing[a]

	Current name	Former names	Function	Motifs
Components of the ~20S editosome	KREPA1	*Tb*MP81, LC-1, band II	Interaction	OB-fold, zinc-finger
	KREPA2	*Tb*MP63, LC-4, band I	Interaction	OB-fold, zinc-finger
	KREPA3	*Tb*MP42, LC-7b, band VI	Interaction*, nuclease	OB-fold, zinc-finger
	KREPA4	*Tb*MP24, LC-10	Interaction	OB-fold?
	KREPA5	*Tb*MP19	Interaction*	OB-fold?
	KREPA6	*Tb*MP18, LC-11, band VII	Interaction*	OB-fold
	KREN1	*Tb*MP90, KREPB1	Deletion endonuclease	RNase III, dsRBM, U1-like
	KREPB2	*Tb*MP67	Endonuclease*	RNase III, dsRBM, U1-like
	KREN2	*Tb*MP61, LC-6a, KREPB3	Insertion endonuclease	RNase III, dsRBM, U1-like
	KREPB4	*Tb*MP46, LC-5	Interaction*	RNase III?,Pumilio, U1-like
	KREPB5	*Tb*MP44, LC-8	Interaction	RNase III?, Pumilio, U1-like
	KREPB6	*Tb*MP49, LC-7c	Interaction*	U1-like
	KREPB7	*Tb*MP47	Interaction*	U1-like
	KREPB8	*Tb*MP41	Interaction*	U1-like
	KREL1	*Tb*MP52, LC-7a, band IV	Ligase	Ligase, tau, K
	KREL2	*Tb*MP48, LC-9, band V	Ligase	Ligase, tau, K
	KREX1	*Tb*MP100, LC-2	ExoUase	5'3' exo, endo/exo/phos
	KREX2	*Tb*MP99, LC-3, band I	ExoUase	5'3' exo, endo/exo/phos
	KRET2	*Tb*MP57, LC-6b	TUTase (editing)	NT, PAP-core, PAP-assoc
	KREH1	*Tb*mHel61p	Helicase	Helicase
Other proteins	KRET1	3' TUTase	TUTase (gRNA)	NT, PAP-core, PAP-assoc, zinc-finger
	TbRGG1	-	Interaction*	RGG
	REAP-1	-	Interaction	21-aa repeat
	RBP16	-	Interaction	Cold shock domain, RGG
	MRP1	*Tb*gBP21, *Lt*p28, *Cfg*BP29	RNA matchmaking	Arg-rich
	MRP2	*Tb*gBP25, *Lt*p26, *Cfg*BP27	RNA matchmaking	Arg-rich

editosome protein genes with the TAG fused to their 3′ end. Mass spectrometric analysis of complexes purified by all these methods led to the identification of 20 kinetoplastid RNA editing proteins (KREPs) and their corresponding genes (Panigrahi et al. 2003b; Stuart et al. 2005; Table 1). Many of these proteins are related pairs, and sequence analysis has identified motifs suggestive of their catalytic activities and/or of RNA or protein interactions (Worthey et al. 2003). Each of six related KREPs, designated KREPA1-6, have a predicted C-terminal oligonucleotide binding (OB) fold, and the three largest (KREPA1-3) have zinc-finger motifs. The functions of these proteins are unknown, but they may play a role in molecular interaction and coordinating the order of catalysis (see Sect. 3). Eight related KREPs that each have an N-terminal U1-like motif were originally designated B1-8; however, two of these proteins, KREPB1 and KREPB3, were identified as kinetoplastid RNA editing endonucleases and renamed KREN1 and KREN2, respectively. KREN1 and KREN2 each have an RNase III motif and a more C-terminal double-stranded RNA binding motif (dsRBM), as does KREPB2. Since KREPB2 also has these characteristic endonuclease motifs, it may also be an RNA endonuclease. The functions of the other five KREPB proteins are uncertain, although KREPB5 is essential for editosome integrity (Wang et al. 2003). The kinetoplastid RNA editing exoUases (KREX1 and 2), and the kinetoplastid RNA editing TUTase KRET2 are components of the ~20S editosome. A related protein, KRET1, is in a separate complex (see Sect. 5.1). Two kinetoplastid RNA editing RNA ligases (KREL1 and 2) are ~20S editosome components, and a single kinetoplastid RNA editing helicase (KREH1) can be associated with ~20S editosomes.

2 Editosome Catalytic Components

2.1 Endonucleases

KREN1 and KREN2 are RNA editing endonucleases specific for deletion and insertion sites, respectively (Carnes et al. 2005; Trotter et al. 2005). Loss of either KREN1 or KREN2 leads to near-total loss of edited mRNAs in vivo, and is lethal. KREN1, KREN2, and KREPB2 are related and all contain a U1-like Zn^{+2} finger motif, an RNase III motif, and a dsRBM sequence. This suggests that KREPB2 may also be a RNA editing endonuclease, but this has not yet been thoroughly assessed. An intriguing possibility is that it may have a specific function such as the stage-specific cleavage of COII pre-mRNA. The U1-like Zn^{+2} finger motif implies that the RNA editing endonucleases are likely involved in complex interactions that require both substrate RNA and other editosome proteins. The RNase III and dsRBM domains are characteristic of known endonucleases, and appear to employ a reaction mechanism similar to that of RNase IIIs, since endonuclease activity is abrogated upon mutation of a key amino acid in the RNase III motif of either KREN1 or KREN2 (Lamontagne et al. 2001; Conrad and Rauhut 2002; Carnes et al. 2005; Trotter et al. 2005).

The editosome accurately edits thousands of editing sites in substrates that have remarkable diversity in the sequences surrounding these editing sites, and only a few sequence limitations exist (Burgess and Stuart 2000; Igo et al. 2002). This diversity reflects the use of numerous gRNAs, many specifying editing of both insertion and deletion sites in the many mRNAs that are edited. Somehow, the endonucleases distinguish between insertion and deletion editing sites in these diverse substrates. These endonuclease activities have distinct biochemical characteristics, in that adenosine nucleotides inhibit insertion cleavage, but stimulate deletion cleavage (Cruz-Reyes et al. 1998). RNA structure, such as a "bulge" in mRNA deletion sites or in gRNA at insertion sites, may contribute to endonuclease discrimination. However, factors such as ~20S editosome composition and interactions among proteins, complexes, and other soluble factors may play a role.

2.2 ExoUases

Following endonucleolytic cleavage at a deletion editing site, unpaired Us are removed by a U-specific exonuclease or exoUase. Two *T. brucei* editosome proteins, KREX1 and KREX2, were originally identified by mass spectrometry, and implicated as possible exoUases because they possess N-terminal 5′→3′ exonuclease (5′3′exo) and C-terminal endo/exo/phosphatase (EEP) motifs. Both KREX1 and KREX2 have exoUase activity in vitro (Kang et al. 2005). RNAi inactivation of KREX1 in vivo results in decreased U removal by cell extracts, and inhibits cell growth. While RNAi inactivation of KREX2 expression does not inhibit cell growth, RNAi knockdown of both KREX1 and KREX2 has greater growth inhibition than does KREX1 RNAi alone (Ernst et al., unpublished data). The specific roles of KREX1 and 2 are unclear. KREX2 is a part of a subcomplex that also contains KREL1 and KREPA2, and this subcomplex has exoUase activity (Schnaufer et al. 2003). In addition, both KREX1 and KREX2 are in complexes purified using TAP-tagged KREN1, but only KREX2 is present in complexes purified using TAP-tagged KREN2 or KREPB2 (Panigrahi et al. 2006). Thus, it seems likely that these two exoUases have distinct functions that might be revealed by further study. Perhaps KREX2 has different substrate requirements and is selectively used for removal of Us from deletion sites with certain characteristics (e.g., following Cs), or it may function to remove excess Us that are added by TUTase at insertion sites. The *Leishmania* KREX2 ortholog lacks the C-terminal EEP motif present in both KREX1 and 2 in *T. brucei* (Aphasizhev et al. 2003a; Worthey et al. 2003). This suggests a more central role in U removal for KREX1, at least in *Leishmania*, and may reflect the less extensive editing and/or different regulation in *Leishmania* compared to *T. brucei*.

2.3 TUTase

Terminal uridylyl transferase (TUTase) catalyzes addition of Us in a gRNA-directed manner after endonucleolytic cleavage at an insertion editing site. Two TUTases,

KRET1 and KRET2, have been identified and they have distinct roles in editing. KRET2 is a component of the ~20S editosome and adds Us to mRNA as specified by gRNA, while KRET1 is a component of a separate complex (see Sect. 5.1) and adds oligo(U) tails to gRNA (Aphasizhev et al. 2002; Ernst et al. 2003). KRET1 and 2 have distinct biochemical characteristics. KRET1 adds many Us to single-stranded RNA in vitro, while KRET2 primarily adds a single U. KRET2 adds the specified number of Us to double-stranded RNAs (dsRNA), whereas KRET1 does not precisely add U to dsRNA. KRET1 and KRET2 are related, and both contain nucleotidyl transferase and poly(A) polymerase (PAP) domains, but the larger KRET1 contains an essential N-terminal C2H2 zinc-finger domain that KRET2 lacks (Aphasizheva et al. 2004). RNAi-mediated knockdown of either KRET1 or KRET2 inhibits growth and reduces the amount of edited mRNAs in vivo, but only loss of KRET2 diminishes in vitro insertion editing (Aphasizhev et al. 2003c). KRET2 interacts with KREPA1, and this interaction enhances TUTase activity (Ernst et al. 2003). In addition, KRET2 is part of a subcomplex that contains KREL2 and KREPA1 (see Sect. 4.1; Schnaufer et al. 2003).

2.4 Ligases

RNA ligases rejoin the mRNA cleavage fragments following U addition or U removal. The ~20S editosome contains two ligases, KREL1 and KREL2, which have been found in deletion and insertion editing subcomplexes, respectively (Cruz-Reyes et al. 2002; Schnaufer et al. 2003). However, the extent to which the physical separation reflects a possible functional separation is unclear (Gao and Simpson 2003). KREL1 and KREL2 are closely related, and both contain five of the six signature motifs of ligases (Schnaufer et al. 2001). KREL1 is essential for editing in vivo, but KREL2 is not, suggesting that KREL1 can compensate for the loss of KREL2, while the reciprocal is not the case (Huang et al. 2001; Schnaufer et al. 2001; Drozdz et al. 2002). KREL1 and KREL2 have distinct biochemical characteristics: they respond differently to ATP and pyrophosphate concentrations (Cruz-Reyes et al. 2002), and differ in tolerance for gaps and overhangs in the substrate (Palazzo et al. 2003). The preferred substrate for the editing ligases is nicked dsRNA, although they can ligate RNAs with overhangs or gaps to various extents (Palazzo et al. 2003). The crystal structure of the large catalytic domain of KREL1 has been determined, and it identifies the residues coordinating ATP binding, a Mg^{+2} atom, and a deep pocket (Deng et al. 2004). While the RNA editing ligases lack the OB-fold motif found in many other ligases, their binding partners, KREPA1 and KREPA2, both possess OB-fold motifs. These binding partners appear to supply the OB-fold function to the editing ligases in *trans* (see Sect. 4.1; Shuman and Schwer 1995; Schnaufer et al. 2001).

2.5 RNA Helicase

The editing process requires that each gRNA be displaced after the editing it specifies is completed, so that the subsequent gRNA can form an anchor duplex with the newly edited region of the mRNA. The necessity of gRNA exchange, as well as other possible requirements for RNA strand separation during editing, such as RNA translocation, indicates a likely involvement of RNA helicase activity in editing. The DEAD box helicase KREH1 appears to fulfill this role in *T. brucei*. KREH1 is not essential, since null mutants can be produced. However, both cell growth and RNA editing are impaired in these mutants, and re-expression of KREH1 in these mutants reverses the editing defect (Missel et al. 1997). Other potential mitochondrial helicases have been identified by in silico analyses, and might be involved in RNA editing (Panigrahi et al. 2003a). KREH1 is found in editosome samples that were purified by biochemical or immunoaffinity methods, but was not detected in several TAP-tag purified editosomes (Panigrahi et al. 2006). This implies that KREH1 has a low-affinity association with the editosome, or it associates with editosome subclasses, which perhaps may represent those at the step of gRNA dissociation.

3   Editosome Structural Components

Eleven of the 20 editosome proteins have no motifs suggestive of catalytic activity, but they do possess motifs suggestive of RNA and/or protein interaction. These proteins might be essential for the structural integrity and perhaps assembly of the editosome, and they are likely to have critical functions in the editing process. They include KREPA1-6 and KREPB4-8. The KREPA family members all possess C-terminal sequences with similarity to the OB-fold motif, suggesting a role in RNA interaction. KREPA1, 2, and 3 also have two more N-terminal C2H2 zinc-finger motifs that are indicative of RNA and/or protein interaction. KREPA3 has been reported to possess nuclease function; however, the activity reported for KREPA3 does not resemble that expected for editing, and its function needs further exploration (Brecht et al. 2005). All members of the KREPB family possess a U1-like zinc-finger motif in the N-terminal region, which is suggestive of RNA and/or protein interaction. The KREN1 and KREN2 endonucleases and KREPB2 all contain the U1-like motif and RNase III motifs, and they are more closely related to each other than to KREPB4-8 members. Unlike the other KREPB members, KREN1, KREN2, and KREPB2 have a more C-terminal dsRBM. These sequence similarities suggest that KREPB2 is a likely endonuclease, although endonuclease activity has not yet been demonstrated. KREPB4 and KREPB5 are more closely related to each other than to the other KREPB proteins, and each has degenerate RNase III and Pumilio RNA binding motifs in their C-terminal region. KREPB5 is essential for editosome structure and stability, since knockdown of its expression results in editosome disruption and disappearance (Wang et al. 2003). KREPB6, KREPB7, and KREPB8 have not been studied to any extent, and they lack recognizable motifs other than the U1-like motif, which suggests a role in molecular interactions.

4   Interactions in the Editosome

4.1   *Separate Insertion and Deletion Subcomplexes*

Some insight has been gained into the interactions between the protein components of the ~20S editosome and its general organization, but much of the architecture of this complex is unknown. Interactions within the editosome are further complicated by studies that indicate editosome heterogeneity in addition to higher-order interactions with other complexes (see Sect. 5). Yeast two-hybrid and immunoprecipitation experiments revealed binary interactions supporting the hypothesis that the ~20S editosome is physically and functionally organized into insertion and deletion editing subcomplexes (Fig. 2; Ernst et al. 2003; Schnaufer et al. 2003). The KRET2 TUTase and KREL2 ligase each interact with KREPA1 physically, and perhaps functionally, linking the U addition and ligation steps in insertion editing. Similarly, the KREX2 exoUase and KREL1 ligase each interact with KREPA2, thus linking

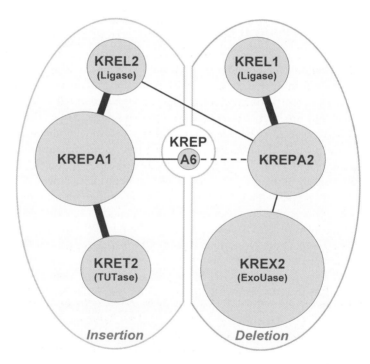

Fig. 2 Schematic of established interactions among editosome proteins. Diameter of *circles* is proportional to the molecular weight of each protein. *Thick lines* represent interactions demonstrated by multiple lines of evidence, including yeast 2-hybrid and co-immunoprecipitation, while *thin lines* represent reproducible yeast 2-hybrid interactions. The *dotted line* between KREPA2 and KREPA6 reflects variability observed with yeast 2-hybrid experiments. *Background lobes* emphasize insertion and deletion subcomplexes

the U removal and ligation steps in deletion editing. These interactions enhance catalytic activity in vitro, suggesting that they occur in vivo. Co-immunoprecipitation studies reveal the same pattern of binary interactions, and indicate that the interactions stimulate the catalytic activities.

The composition of heterotrimeric subcomplexes recovered from cells that had been engineered to conditionally express TAP-tagged KREL1 or KREL2 reinforces binary interaction data (Schnaufer et al. 2003). In each cell line, the tagged ligase was incorporated into ~20S editosomes with compositions similar to those purified biochemically, and these editosomes could perform pre-cleaved insertion and deletion editing. However, in each cell line the TAP-tagged ligase was incorporated into a different, smaller (~5–10S) complex that had a composition that mirrored the yeast two-hybrid and immunoprecipitation results. Tagged KREL1 was in a complex with both KREPA2 and KREX2, while tagged KREL2 was in a complex with both KREPA1 and KRET2. The former complex catalyzed U removal and ligation, and the latter catalyzed U addition and ligation. These results corroborate the interaction data, and provide insights into how these proteins might work together. Schnaufer and colleagues suggest that KREPA1 and KREPA2 provide the missing ligase motif (the OB-fold) in *trans*, which functions in substrate recognition and binding, as shown for DNA ligases and RNA capping enzymes (Schnaufer et al. 2003). They hypothesize that this arrangement provides for U addition or removal occurring prior to ligation via OB-fold conformation changes similar to those that occur in the DNA ligases and capping enzymes. Perhaps the protein interactions in each heterotrimer are strong in order to tightly associate the proteins into an "enzymatic whole". This might explain why these binary interactions were preferentially detected in the two-hybrid and immunoprecipitation studies. These studies indicate that the ~20S editosome is structurally and functionally segregated into insertion and deletion subcomplexes. However, this aspect is not quite that simple, as is discussed in Sections 4.2 to 4.4.

4.2   Multifaceted Interactions Within the Editosome

Multiple interactions among proteins and/or RNA may be required for stable association of some editosome proteins in vivo. KREPA6 shows weak interaction with KREPA1 and KREPA2 in yeast two-hybrid studies, and gene expression knockdown studies indicate that KREPB5 and KREPA4 associate with both the insertion and deletion subcomplexes, and that they are essential for ~20S editosome integrity and stability (Wang et al. 2003; Salavati et al. 2006). Such molecular interactions are consistent with their motifs. Indeed, eight editosome proteins have U1-like motifs and may function in molecular interactions. The archetypal U1-like sequence, found in human snRNP protein C, requires the presence of one or more of the U1 snRNP proteins in order to bind to U1 snRNA

(Gunnewiek et al. 1995). Since numerous editosome proteins have U1-like motifs, they may mediate similar complex interactions that require both RNA and protein. However, neither pre-edited RNA nor gRNA are required to assemble editosomes that catalyze editing in vitro, as shown by mutants that lack genes for these RNAs but contain functional editosomes (Domingo et al. 2003). Nevertheless, multiple proteins may be required for specific RNA binding, and reciprocally, RNA binding may stabilize associations of proteins with the editosome.

4.3   Functionally and Compositionally Distinct ~20S Editosomes

Studies of complexes from cells expressing tagged KREN1, KREN2, and KREPB2 RNase III proteins suggest that the cells contain editosomes that differ in composition (Fig. 3; Panigrahi et al. 2006). In contrast to biochemically purified editosomes, which have been used to detect all known editosome proteins, TAP purified editosomes possess particular subsets of editosome proteins, depending on the protein that is tagged. Each of these purified editosomes contains KREPA1-4, KREPA6, KREPB4 and 5, KRET2, KREX2, KREL1, and KREL2. However, KREN1, KREN2, and KREPB2 were mutually exclusive in complexes in which one of these proteins was tagged. KREX1 was found only in the KREN1-TAP complexes, as was KREPB8, while KREPB7 was found only in KREN2-TAP complexes. KREPB2-TAP complexes also apparently lack KREPA5. The distinct composition of these editosomes is reflected in their functional differences. KREN1-TAP complexes cleave only deletion editing sites, while KREN2-TAP complexes cleave only insertion editing sites. The exclusive presence of KREX1 in KREN1-TAP editosomes suggests that KREX1 might be more immediately involved in U removal following cleavage. These results indicate that editosomes can be somewhat heterogeneous in composition in vivo, implying that these different editosomes may interact to catalyze RNA editing, or that editosome composition is dynamic, with proteins (e.g., endonucleases) shuttling in and out of the complex.

4.4   ~40S Editosome Complexes

Although most in vitro experiments to date have dealt with the ~20S editosome, editosome proteins can also be detected in larger, ~40S complexes that contain significant amounts of gRNA, pre-edited mRNA, and partially edited mRNA (Pollard et al. 1992; Corell et al. 1996). Currently, the ~40S complex is poorly understood, and little is known about its role in RNA editing. The associations between the components of the ~40S complex are apparently weaker than those in the ~20S complex, as the ~40S complex is more sensitive to increasing ionic

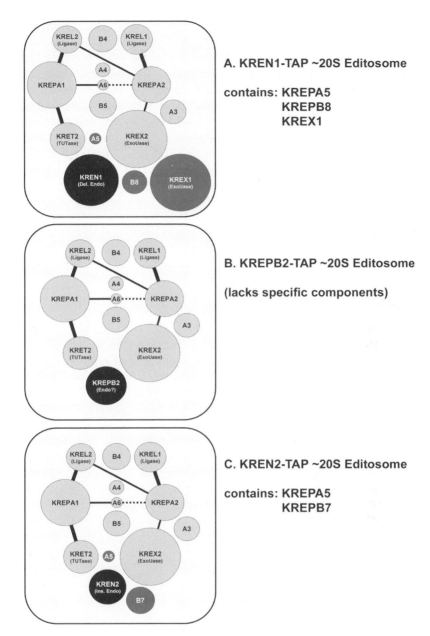

Fig. 3 Model for multiple ~20S editosome compositions based on TAP-tag experiments. Following tandem affinity purification, compositionally different ~20S editosomes are isolated depending upon whether KREN1 (**A**), KREPB2 (**B**), or KREN2 (**C**) is TAP-tagged. TAP-tagged proteins are *black*, while specifically associating proteins are *dark grey*. Diameter of *circles* is proportional to the molecular weight of each protein. Core editosome proteins, including those with known binary interactions shown in Fig. 2, are consistently found in each editosome. However, KREPB8 and KREX1 are found only in KREN1-TAP complexes, and KREPB7 only in KREN2-TAP complexes. Although this model shows three separate ~20S editosomes, this is not meant to imply that dynamic exchange does not occur. The proteins specifically associated with KREN1, KREPB2, or KREN2 might shuttle in and out of "editosome cores" consisting of the *light grey* proteins

strength (Pollard et al. 1992). The sedimentation of the ~40S complex changes depending upon the extent of editing in mRNAs in the complex, which may be due to the increased size of the mRNA itself, or other factors that associate with edited RNA (Pollard et al. 1992). Because larger RNAs correlate with larger ~40S complexes, one might surmise that the difference in sedimentation between ~20S and ~40S complexes is due to the additional RNAs, proteins that associate with these RNAs, or both. Comparison of the ~20S and ~40S complexes indicates that the ~20S complex has a significantly greater capacity for in vitro editing of exogenously added RNAs, suggesting that RNA already in the ~40S complex prevents editing of subsequently added RNAs (Corell et al. 1996). Distribution of pre-edited and partially edited mRNAs is also distinct between ~20S and ~40S complexes, as increasingly edited mRNAs correlate with larger complexes.

5   Other Complexes and Proteins

The ~20S editosome complex contains the activities for endonucleolytic cleavage, U addition and U removal, and RNA ligation. Several proteins that are not stable components of the ~20S editosome have roles in the overall process of editing, either directly or indirectly. Some of these proteins are found in separate complexes of their own, and these complexes may interact with the ~20S editosome to form higher-order complexes. These include the KRET1 complex that adds the Us to the gRNAs, and the MRP complex that has matchmaking activity, i.e., annealing two RNAs and releasing them. These complexes, and perhaps others, can affect editing even though they do not appear to participate in the central catalytic steps that process the mRNA directly. Other proteins, including RBP16, REAP-1, and TbRGG1, may have roles associated with RNA editing. The following discussion centers on the roles played by those proteins that are extra-editosomal factors.

5.1   KRET1

KRET1, which is related to the insertion editing TUTase KRET2 (see Sect. 2.3), is present in a complex that sediments at ~10S in glycerol gradients, and catalyzes addition of the 3′ oligo(U) tail to gRNAs (Aphasizhev et al. 2002). Recombinant KRET1 can oligomerize in vitro, suggesting that ~10S KRET1 complexes might contain multiple KRET1 proteins. KRET1 complexes also appear to associate with the ~20S editosome in an RNase-sensitive manner, leading to the speculation that the KRET1 complex may also function in loading gRNA onto the ~20S editosome (Aphasizhev et al. 2002). Loss of KRET1 in vivo decreases viability and RNA editing, and results in gRNAs with shorter oligo(U) tails (Aphasizhev et al. 2002, 2003c). This loss of viability suggests that the oligo(U) tail is essential in vivo, although it is not required in vitro (Burgess et al. 1999). In addition to its role in editing, KRET1 is required for the UTP-dependent degradation of polyadenylated RNAs in *T. brucei* mitochondria (Ryan and Read 2005). Like KRET1, other proteins involved in editing may have additional functions outside editing.

5.2 MRPs

The mitochondrial RNA binding proteins (MRP1 and MRP2) are related to each other and both have high affinity for RNAs. MRP1 and MRP2 form stable heterotetramers, and they catalyze annealing of complementary RNA (Blom et al. 2001; Aphasizhev et al. 2003b). The MRP complex is separate from the ~20S editosome, but like the KRET1 complex, it also appears to associate with the ~20S complex in an RNase-sensitive manner (Aphasizhev et al. 2003b). Although MRP1 binds various RNAs, immunoprecipitation of either MRP1 or MRP2 co-precipitates gRNAs (Allen et al. 1998; Aphasizhev et al. 2003b). RNAi-mediated knockdown of either MRP1 or MRP2 leads to the loss of both proteins, inhibits growth, and affects the relative abundance of edited and unedited RNAs (Vondruskova et al. 2005). Because they bind gRNAs and possess RNA matchmaking activity, the MRPs could play a role in associating gRNA with pre-mRNA. MRP association with ~20S editosomes is RNA-mediated, and might reflect interaction with mRNAs that are associated with ~20S editosomes. The effect of MRP1 and MRP2 knockdown on the relative abundance of various edited and unedited mRNAs suggests that they may play a role in regulating RNA editing.

5.3 RBP16

An RNA binding protein with a mass of ~16 kDa (RBP16) also affects editing, and like MRP1 and MRP2, may be regulatory. RBP16 has an N-terminal cold shock domain and C-terminal region rich in arginine and glycine, which implies a role in post-transcriptional gene regulation (Hayman and Read 1999). RBP16 binds gRNAs (with an affinity for oligo(U)), rRNAs, and mRNAs (Hayman and Read 1999; Pelletier and Read 2003). Although RBP16 apparently does not associate with ~20S editosome, it does interact with other proteins, one of which (p22) modulates RBP16 affinity for gRNA (Hayman et al. 2001). RNAi-mediated knockdown of RBP16 in PF *T. brucei* dramatically decreases edited CYb mRNA, but does not reduce pre-edited CYb, and importantly does not affect the amount of edited A6, COII, or MURF2 mRNAs (Pelletier and Read 2003). Furthermore, RBP16 knockdown also decreases the levels of COI and ND4 mRNAs, which do not undergo editing. Thus, the role of RBP16 is not limited to edited RNAs. While the precise function of RBP16 is unclear, it ultimately impacts the levels of edited RNA.

5.4 REAP-1 and TbRGG1

RNA editing associated protein 1 (REAP-1) is a ~45 kDa protein that is in 35–40S complexes, and overlaps the sedimentation profiles of TUTase and RNA ligase (Madison-Antenucci et al. 1998). Antibodies specific for REAP-1 inhibit in vitro editing, suggesting a role associated with editing (Madison-Antenucci et al. 1998). Studies indicating preferential binding of REAP-1 to pre-edited mRNAs led to the suggestion that it may play a role in loading these substrates onto the editosome (Madison-Antenucci and Hajduk 2001). However, these suggested roles have not yet been confirmed experimentally. TbRGG1 is ~75 kDa mitochondrial protein named for its five Arg-Gly-Gly (RGG) repeats that are suggestive of RNA binding, and has been shown to have a preference for oligo(U) (Vanhamme et al. 1998). TbRGG1 co-sediments with in vitro deletion editing activity in glycerol gradients, but direct interaction with the editosome has not been demonstrated, nor has a role in RNA editing.

6 RNA Editing as a Drug Target

RNA editing may be a novel target for drug development, since hosts lack editing and editing is normally essential in the pathogenic bloodstream stage of *T. brucei*, as shown by lethality resulting from the loss of KREL1, KREPB5, KREN1, or KREN2 (Schnaufer et al. 2001; Wang et al. 2003; Carnes et al. 2005; Trotter et al. 2005). However, these pathogens can bypass the requirement for RNA editing. Dyskinetoplastic (DK) strains are viable although they cannot edit due to loss of much or all of their mitochondrial DNA. Compensatory mutations in respiratory proteins appear to allow this bypass (Schnaufer et al. 2005). The number of compensatory mutations that are required, and the frequency with which they occur may affect the utility of editing as a drug target.

Acknowledgements The authors would like to thank Nancy Lewis Ernst and Achim Schnaufer for helpful discussions during the writing of this review. This work was supported by NIH grants GM042188 (to K. Stuart) that extended work supported by AI014102 (to K. Stuart). J. Carnes was supported by NIH AI007509 (Pathobiology Training Grant, University of Washington). Additional support was provided by the Economic Development Administration – US Department of Commerce, and the M.J. Murdock Charitable Trust.

References

Abraham JM, Feagin JE, Stuart K (1988) Characterization of cytochrome c oxidase III transcripts that are edited only in the 3' region. Cell 55:267–272

Allen TE, Heidmann S, Reed R, Myler PJ, Göringer HU, Stuart KD (1998) Association of guide RNA binding protein gBP21 with active RNA editing complexes in *Trypanosoma brucei*. Mol Cell Biol 18:6014–6022

Aphasizhev R, Sbicego S, Peris M, Jang SH, Aphasizheva I, Simpson AM, Rivlin A, Simpson L (2002) Trypanosome mitochondrial 3' terminal uridylyl transferase (TUTase): the key enzyme in U-insertion/deletion RNA editing. Cell 108:637–648

Aphasizhev R, Aphasizheva I, Nelson RE, Gao G, Simpson AM, Kang X, Falick AM, Sbicego S, Simpson L (2003a) Isolation of a U-insertion/deletion editing complex from *Leishmania tarentolae* mitochondria. EMBO J 22:913–924

Aphasizhev R, Aphasizheva I, Nelson RE, Simpson L (2003b) A 100-kD complex of two RNA-binding proteins from mitochondria of *Leishmania tarentolae* catalyzes RNA annealing and interacts with several RNA editing components. RNA 9:62–76

Aphasizhev R, Aphasizheva I, Simpson L (2003c) A tale of two TUTases. Proc Natl Acad Sci USA 100:10617–10622

Aphasizheva I, Aphasizhev R, Simpson L (2004) RNA-editing terminal uridylyl transferase 1: identification of functional domains by mutational analysis. J Biol Chem 279:24123–24130

Blom D, van den Berg M, Breek CKD, Speijer D, Muijsers AO, Benne R (2001) Cloning and characterization of two guide RNA-binding proteins from mitochondria of *Crithidia fasciculata:* gBP27, a novel protein, and gBP29, the orthologue of *Trypanosoma brucei* gBP21. Nucleic Acids Res 29:2950–2962

Blum B, Simpson L (1990) Guide RNAs found in kinetoplastid mitochondria contain a non-encoded oligo-[U] 3' tail. Cell 62:391–397

Brecht M, Niemann M, Schlüter E, Müller UF, Stuart K, Göringer HU (2005) TbMP42, a protein component of the RNA editing complex in African trypanosomes has endo-exoribonuclease activity. Mol Cell 17:621–630

Burgess MLK, Stuart K (2000) Sequence bias in edited kinetoplastid RNAs. RNA 6:1492–1497

Burgess MLK, Heidmann S, Stuart KD (1999) Kinetoplastid RNA editing does not require the terminal 3' hydroxyl of guide RNA but modifications to the guide RNA terminus can inhibit *in vitro* U insertion. RNA 5:883–892

Carnes J, Trotter JR, Ernst NL, Steinberg AG, Stuart K (2005) An essential RNase III insertion editing endonuclease in *Trypanosoma brucei*. Proc Natl Acad Sci USA 102:16614–16619

Conrad C, Rauhut R (2002) Ribonuclease III: new sense from nuisance. Int J Biochem Cell Biol 34:116–129

Corell RA, Feagin JE, Riley GR, Strickland T, Guderian JA, Myler PJ, Stuart K (1993) *Trypanosoma brucei* minicircles encode multiple guide RNAs which can direct editing of extensively overlapping sequences. Nucleic Acids Res 21:4313–4320

Corell RA, Read LK, Riley GR, Nellissery JK, Allen TE, Kable ML, Wachal MD, Seiwert SD, Myler PJ, Stuart KD (1996) Complexes from *Trypanosoma brucei* that exhibit deletion editing and other editing-associated properties. Mol Cell Biol 16:1410–1418

Cruz-Reyes J, Rusché LN, Piller KJ, Sollner-Webb B (1998) *T. brucei* RNA editing: adenosine nucleotides inversely affect U-deletion and U-insertion reactions at mRNA cleavage. Mol Cell 1:401–409

Cruz-Reyes J, Zhelonkina AG, Huang CE, Sollner-Webb B (2002) Distinct functions of two RNA ligases in active *Trypanosoma brucei* RNA editing complexes. Mol Cell Biol 22:4652–4660

Decker CJ, Sollner-Webb B (1990) RNA editing involves indiscriminate U changes throughout precisely defined editing domains. Cell 61:1001–1011

Deng J, Schnaufer A, Salavati R, Stuart KD, Hol WGJ (2004) High resolution crystal structure of a key editosome enzyme from *Trypanosoma brucei*: RNA editing ligase 1. J Mol Biol 343:601–613

Domingo GJ, Palazzo SS, Wang B, Panicucci B, Salavati R, Stuart KD (2003) Dyskinetoplastic *Trypanosoma brucei* contain functional editing complexes. Eukaryot Cell 2:569–577

Drozdz M, Palazzo SS, Salavati R, O'Rear J, Clayton C, Stuart K (2002) TbMP81 is required for RNA editing in *Trypanosoma brucei*. EMBO J 21:1791–1799

Ernst NL, Panicucci B, Igo RP Jr, Panigrahi AK, Salavati R, Stuart K (2003) TbMP57 is a 3' terminal uridylyl transferase (TUTase) of the *Trypanosoma brucei* editosome. Mol Cell 11:1525–1536

Gao G, Simpson L (2003) Is the *Trypanosoma brucei* REL1 RNA ligase specific for U-deletion RNA editing, and is the REL2 RNA ligase specific for U-insertion editing? J Biol Chem 278:27570–27574

Gunnewiek JM, van Aarssen Y, Wassenaar R, Legrain P, van Venrooij WJ, Nelissen RL (1995) Homodimerization of the human U1 snRNP-specific protein C. Nucleic Acids Res 23:4864–4871

Hayman ML, Read LK (1999) *Trypanosoma brucei* RBP16 is a mitochondrial Y-box family protein with guide RNA binding activity. J Biol Chem 274:12067–12074

Hayman ML, Miller MM, Chandler DM, Goulah CC, Read LK (2001) The trypanosome homolog of human p32 interacts with RBP16 and stimulates its gRNA binding activity. Nucleic Acids Res 29:5216–5225

Horváth A, Berry EA, Maslov DA (2000) Translation of the edited mRNA for cytochrome b in trypanosome mitochondria. Science 287:1639–1640

Huang CE, Cruz-Reyes J, Zhelonkina AG, O'Hearn S, Wirtz E, Sollner-Webb B (2001) Roles for ligases in the RNA editing complex of *Trypanosoma brucei*: band IV is needed for U-deletion and RNA repair. EMBO J 20:4694–4703

Igo RP Jr, Lawson SD, Stuart K (2002) RNA sequence and base pairing effects on insertion editing in *Trypanosoma brucei*. Mol Cell Biol 22:1567–1576

Kable ML, Seiwert SD, Heidmann S, Stuart K (1996) RNA editing: a mechanism for gRNA-specified uridylate insertion into precursor mRNA. Science 273:1189–1195

Kang X, Rogers K, Gao G, Falick AM, Zhou S, Simpson L (2005) Reconstitution of uridine-deletion precleaved RNA editing with two recombinant enzymes. Proc Natl Acad Sci USA 102:1017–1022

Koslowsky DJ, Bhat GJ, Read LK, Stuart K (1991) Cycles of progressive realignment of gRNA with mRNA in RNA editing. Cell 67:537–546

Lamontagne B, Larose S, Boulanger J, Elela SA (2001) The RNase III family: a conserved structure and expanding functions in eukaryotic dsRNA metabolism. Curr Issues Mol Biol 3:71–78

Leung SS, Koslowsky DJ (1999) Mapping contacts between gRNA and mRNA in the trypanosome RNA editing. Nucleic Acids Res 27:778–787

Leung SS, Koslowsky DJ (2001) Interactions of mRNAs and gRNAs involved in trypanosome mitochondrial RNA editing: structure probing of an mRNA bound to its cognate gRNA. RNA 7:1803–1816

Madison-Antenucci S, Hajduk S (2001) RNA editing-associated protein 1 is an RNA binding protein with specificity for preedited mRNA. Mol Cell 7:879–886

Madison-Antenucci S, Sabatini RS, Pollard VW, Hajduk SL (1998) Kinetoplastid RNA-editing-associated protein 1 (REAP-1): a novel editing complex protein with repetitive domains. EMBO J 17:6368–6376

Missel A, Souza AE, Norskau G, Göringer HU (1997) Disruption of a gene encoding a novel mitochondrial DEAD-box protein in *Trypanosoma brucei* affects edited mRNAs. Mol Cell Biol 17:4895–4903

Palazzo SS, Panigrahi AK, Igo RP Jr, Salavati R, Stuart K (2003) Kinetoplastid RNA editing ligases: complex association, characterization, and substrate requirements. Mol Biochem Parasitol 127:161–167

Panigrahi AK, Allen TE, Haynes PA, Gygi SP, Stuart K (2003a) Mass spectrometric analysis of the editosome and other multiprotein complexes in *Trypanosoma brucei*. J Am Soc Mass Spectrom 14:728–735

Panigrahi AK, Schnaufer A, Ernst NL, Wang B, Carmean N, Salavati R, Stuart K (2003b) Identification of novel components of *Trypanosoma brucei* editosomes. RNA 9:484–492

Panigrahi AK, Ernst NL, Domingo GJ, Fleck M, Salavati R, Stuart K (2006) Compositionally and functionally distinct editosomes in *Trypanosoma brucei*. RNA 12:1038–1049

Pelletier M, Read LK (2003) RBP16 is a multifunctional gene regulatory protein involved in editing and stabilization of specific mitochondrial mRNAs in *Trypanosoma brucei*. RNA 9:457–468

Pollard VW, Harris ME, Hajduk SL (1992) Native mRNA editing complexes from *Trypanosoma brucei* mitochondria. EMBO J 11:4429–4438

Read LK, Myler PJ, Stuart K (1992) Extensive editing of both processed and preprocessed maxicircle CR6 transcripts in *Trypanosoma brucei*. J Biol Chem 267:1123–1128

Rigaut G, Shevchenko A, Rutz B, Wilm M, Mann M, Seraphin B (1999) A generic protein purification method for protein complex characterization and proteome exploration. Nat Biotechnol 17:1030–1032

Riley GR, Myler PJ, Stuart K (1995) Quantitation of RNA editing substrates, products and potential intermediates: implication for developmental regulation. Nucleic Acids Res 23:708–712

Ryan CM, Read LK (2005) UTP-dependent turnover of *Trypanosoma brucei* mitochondrial mRNA requires UTP polymerization and involves the RET1 TUTase. RNA 11:763–773

Salavati R, Ernst N, O'Rear J, Gilliam T, Tarun S, Stuart K (2006) KREPA4, an RNA binding protein essential for editosome integrity and survival of *Trypanosoma brucei*. RNA 12:819–831

Schnaufer A, Panigrahi AK, Panicucci B, Igo RP Jr, Salavati R, Stuart K (2001) An RNA ligase essential for RNA editing and survival of the bloodstream form of *Trypanosoma brucei*. Science 291:2159–2162

Schnaufer A, Ernst N, O'Rear J, Salavati R, Stuart K (2003) Separate insertion and deletion subcomplexes of the *Trypanosoma brucei* RNA editing complex. Mol Cell 12:307–319

Schnaufer A, Clark-Walker GD, Steinberg AG, Stuart K (2005) The F1-ATP synthase complex in bloodstream stage trypanosomes has an unusual and essential function. EMBO J 24:4029–4040

Seiwert SD, Heidmann S, Stuart K (1996) Direct visualization of uridylate deletion in vitro suggests a mechanism for kinetoplastid RNA editing. Cell 84:831–841

Shu H-H, Stuart KD (1994) Mitochondrial transcripts are processed but are not edited normally in *Trypanosoma equiperdum* (ATCC 30019) which has kDNA sequence deletion and duplication. Nucleic Acids Res 22:1696–1700

Shuman S, Schwer B (1995) RNA capping enzyme and DNA ligase: a superfamily of covalent nucleotidyl transferases. Mol Microbiol 17:405–410

Stuart K, Allen TE, Heidmann S, Seiwert SD (1997) RNA Editing in kinetoplastid protozoa. Microbiol Mol Biol Rev 61:105–120

Stuart KD, Schnaufer A, Ernst NL, Panigrahi AK (2005) Complex management: RNA editing in trypanosomes. Trends Biochem Sci 30:97–105

Sturm NR, Simpson L (1990) Partially edited mRNAs for cytochrome b and subunit III of cytochrome oxidase from *Leishmania tarentolae* mitochondria: RNA editing intermediates. Cell 61:871–878

Trotter JR, Ernst NL, Carnes J, Panicucci B, Stuart K (2005) A deletion site editing endonuclease in *Trypanosoma brucei*. Mol Cell 20:403–412

Vanhamme L, Perez-Morga D, Marchal C, Speijer D, Lambert L, Geuskens M, Alexandre S, Ismaïli N, Göringer U, Benne R, Pays E (1998) *Trypanosoma brucei* TBRGG1, a mitochondrial oligo(u)-binding protein that co-localizes with an in vitro RNA editing activity. J Biol Chem 273(34):21825–21833

Vondruskova E, Van den Burg J, Zikova A, Ernst NL, Stuart K, Benne R, Lukes J (2005) RNA interference analyses suggest a transcript-specific regulatory role for MRP1 and MRP2 in RNA editing and other RNA processing in *Trypanosoma brucei*. J Biol Chem 280:2429–2438

Wang B, Ernst NL, Palazzo SS, Panigrahi AK, Salavati R, Stuart K (2003) TbMP44 is essential for RNA editing and structural integrity of the editosome in *Trypanosoma brucei* . Eukaryot Cell 2:578–587

Worthey EA, Schnaufer A, Mian IS, Stuart K, Salavati R (2003) Comparative analysis of editosome proteins in trypanosomatids. Nucleic Acids Res 31:6392–6408

RNA Editing Accessory Factors – the Example of mHel61p

H. Ulrich Göringer (✉), Michael Brecht, Cordula Böhm, and Elisabeth Kruse

> *"Nature uses only the longest threads to weave her patterns, so each small piece of her fabric reveals the organization of the entire tapestry."*
>
> Richard Feynman
> *The character of physical law.*

Abstract The majority of mitochondrial pre-messenger RNAs in kinetoplastid protozoa are substrates of a U nucleotide-specific, insertion/deletion-type RNA editing reaction. The process converts nonfunctional pre-mRNAs into translatable molecules, and can generate protein diversity by alternative editing. A high molecular mass enzyme complex, the editosome, catalyzes the reaction. Editosomes provide a molecular platform for the individual catalytic steps of the reaction cycle. While the molecular composition of the editosome has been studied in detail, dynamic aspects of the reaction have by and large been ignored. Here, we focus on accessory proteins that bind to the editosome only at defined steps of the reaction cycle, thereby modulating the structure and function of the catalytic machinery. As an example, we concentrate on the mitochondrial DExH/D protein mHel61p, a putative RNA helicase and/or RNPase. We summarize the current structural, genetic and biochemical knowledge on mHel61p, and provide an outlook onto dynamic processes of the editing reaction.

Genetics, Darmstadt University of Technology, Schnittspahnstr. 10, 64287 Darmstadt, Germany;
goringer@hrzpub.tu-darmstadt.de

H.U. Göringer (ed.), *RNA Editing. Nucleic Acids and Molecular Biology 20*
© Springer-Verlag Berlin Heidelberg 2008

1   Introduction

RNA editing of mitochondrial pre-messenger RNAs (pre-mRNAs) in kinetoplastid Protozoa is a unique form of posttranscriptional gene transcript maturation. The process is characterized by the site-specific insertion and deletion of exclusively uridylate residues, thereby creating functional mRNA molecules for translation (Benne et al. 1986; for a recent review, see Stuart et al. 2005). RNA editing relies on small, metabolically stable RNA molecules known as guide RNAs (gRNAs), which act as "quasi" templates in the process (Blum et al. 1990). Guide RNAs initiate the processing reaction by the formation of a short, intermolecular duplex structure with the pre-mRNA, located proximal to the sequence domain to be edited. Bordering this helical segment, unpaired gRNA nucleotides then specify, via base pairing, the U-insertion event with free UTP as the substrate. Non-base-paired uridylates in the pre-mRNA become deleted (Seiwert and Stuart 1994; Frech and Simpson 1996; Kable et al. 1996; Seiwert et al. 1996). By utilizing different gRNAs, alternative editing events can take place, thereby generating protein diversity within the mitochondria of the parasites (Ochsenreiter and Hajduk 2006). The entire reaction series is mediated by high molecular mass protein complexes, known as editosomes (Pollard et al. 1992; Corell et al. 1996; Rusché et al. 1997; Stuart et al. 2005). Editosomes function as a reaction platform for the pre-edited mRNA/gRNA hybrid molecules, and catalyze the processing reaction in a series of enzyme-driven steps. In its most basic form, this involves endo/exonuclease activities, terminal uridylyl transferases (TUTases), and RNA ligase-type catalysts.

Over the last years, our knowledge of the inventory of the editosome has significantly increased (see chapter by Carnes and Stuart, this volume). Depending on the enrichment protocol, active RNA editing complexes contain as little as seven (Rusché et al. 1997), 13 (Aphasizhev et al. 2003), or up to 20 polypeptides (Panigrahi et al. 2001). Protein candidates for every step of the minimal reaction cycle have been identified (Aphasizhev and Simpson 2001; McManus et al. 2001; Rusché et al. 2001; Aphasizhev et al. 2002; Brecht et al. 2005; Carnes et al. 2005; Trotter et al. 2005), thereby confirming the general features of the above-described enzyme-driven reaction model. The complexes have an apparent hydrodynamic size in the range of 20 Svedberg units, and are capable of conducting in vitro RNA editing when supplied with cognate gRNA/pre-mRNA hybrid molecules. However, the inability of the 20S complexes to carry out more than one round of RNA editing, as well as the association of endogenous pre-mRNA molecules with larger complexes ($\leq$40S; Pollard et al. 1992; Madison-Antenucci et al. 2002) suggest that these high molecular mass RNA-containing complexes catalyze the editing reaction in vivo. Unfortunately, no detailed structural picture of these complexes is known today. Other critical information is lacking as well. This includes the questions of whether the editing reaction is distributive or processive, whether the editosome has one or multiple reaction centers, and whether the U-insertion and U-deletion reactions are catalyzed by separate editing complexes (Gao and Simpson 2003; Aphasizhev et al. 2003; Schnaufer et al. 2003; Panigrahi et al. 2006).

Similarly, no knowledge as to the dynamic characteristics of the reaction exists. In analogy to other biochemical processes that are catalyzed by macromolecular protein or ribonucleoprotein (RNP) complexes, it is likely that the editing reaction is highly dynamic. The participating RNA and protein molecules possibly undergo defined conformational changes not only during the assembly and disassembly of the editosomal complex, but also as the reaction proceeds and terminates. Especially within the context of the catalytic center(s) of the editosome, it is almost certain that dynamic rearrangements must occur. The catalytic "core" must provide access for several enzyme activities (see above), and the pre-mRNA and gRNA molecules alter their structure during the course of the reaction. This likely requires several repositioning steps for both the editosomal proteins and the pre-mRNA/gRNA hybrid molecules. These dynamic steps may be catalyzed either by polypeptides of the editing complex with as yet unassigned functions, or by accessory factors, which temporarily bind to the editosome in order to execute a defined reaction step. In analogy to other protein- and RNP-complex-driven biochemical reactions, there are many opportunities for accessory proteins to modulate the editing reaction cycle (Fig. 1). Several RNA editing auxiliary proteins have already been described. These include REAP1, a protein that interacts with, and stabilizes pre-edited mRNA (Madison-Antenucci and Hajduk 2001; Hans et al. 2007), TbRGG1, a polypeptide that can bind to the 3′ oligo(U) extensions of gRNAs (Vanhamme et al. 1998), and the RNA-binding protein RBP16 (Miller and Read 2003). Also included are the matchmaking-type gRNA/mRNA annealing factors gBP21 and gBP25 (Köller et al. 1997; Allen et al. 1998; Blom et al. 2001; Müller et al. 2001; Müller and Göringer 2002; Schumacher et al. 2006; for a detailed discussion, see the chapter by Homann, this volume).

The purpose of this article is to elaborate on the concept of "accessory protein factors" that are part of the editing cycle, and contribute to dynamic steps of the reaction. A special focus will be put on the mitochondrial DExH/D-box protein mHEL61p, a putative RNA helicase/RNPase. We summarize what is known about mHEL61p, and provide an outlook onto dynamic aspects of the RNA editing reaction in African trypanosomes.

2   mHel61p – a Protein of the DExH/D-Box Protein Family

The concept of structurally dynamic steps within the RNA editing reaction cycle was put forward very early, based on the fact that some pre-mRNAs are edited throughout their entire length. The reaction proceeds with a general 3′ to 5′ directionality (on the pre-edited mRNA), and requires the sequential base pairing of multiple gRNA molecules (Maslov and Simpson 1992). To allow the ordered interplay of successive gRNAs, which often partially overlap, the participation of an RNA chaperone function in the form of an RNA helicase was postulated (Missel and Göringer 1994). In fact, a mitochondrial RNA unwinding activity was identified in *Trypanosoma brucei*, displaying all characteristic features of RNA helicase

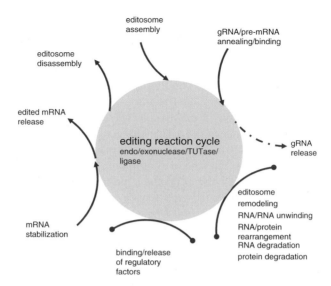

Fig. 1 Summary of biochemical processes that, in combination with the U-specific deletion/insertion RNA editing reaction, could be regulated by accessory (protein) factors

enzymes: the activity was able to unwind tailed duplex RNAs in a nucleoside-triphosphate (NTP)- and Mg^{2+}-cation-dependent manner, but failed to unwind double-stranded DNA (Missel and Göringer 1994).

The majority of RNA helicases or unwindases belong to the superfamily of DExH/D proteins. These enzymes are known to modulate RNA structure in an ATP-dependent way (Fuller-Pace 1994). Members of this protein family are present in all forms of cellular life, and have also been identified in viruses. The polypeptides are highly conserved, and share nine characteristic amino acid signature motifs (reviewed in Linder 2006; Fig. 2A). Sequence motif II often has the form of DEAD, DEAH or DExH, which provides the names for the three subgroups of the protein superfamily. DExH/D-box proteins have been shown to play crucial roles in almost all cellular processes that require RNA/RNA and RNA/protein interactions, including nuclear and mitochondrial pre-mRNA splicing, ribosomal RNA processing, RNA degradation and stabilization, translation initiation, and ribosome assembly (Schmid and Linder 1992; Linder 2006). Based on the sequence homology of DExH/D proteins to DNA helicases, and the fact that some prototypic members of the family exhibit RNA helicase activity in vitro, it has been proposed that DExH/D proteins act predominantly as ATP-dependent RNA helicases. However, more recent evidence suggests that the proteins act as multifunctional ATP-driven motors or switches. They are able to regulate multiple reaction steps that require the formation of optimal RNA structures through local RNA unwinding, or they function as RNPases by catalyzing association and/or dissociation events within RNA/protein complexes (for a recent review, see Jankowsky and Bowers 2006).

(A)

(B)

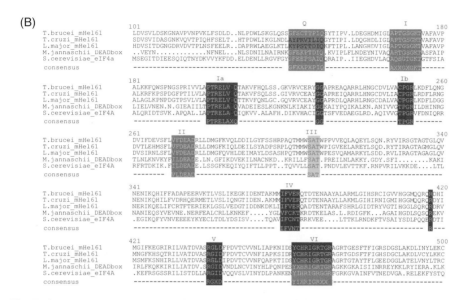

Fig. 2 A Schematic representation of the nine conserved amino acid signatures of DExH/D-box family proteins. The different colors indicate the various functions of the individual motifs. *Red* ATP binding and hydrolysis, *purple* RNA binding, *magenta* RNA unwinding, *blue* conserved, but no function assigned. **B** Sequence alignment of the three kinetoplastid mHel61p sequences from *Trypanosoma brucei*, *Trypanosoma cruzi*, and *Leishmania major*, compared to the prototypic DExH/D-box proteins eIF4a from *S. cerevisiae* and mjDEAD from *M. jannaschii*. Consensus motifs are color coded as in **A**. *X* Variable amino acid, *N* amino-terminus, *C* carboxy-terminus. Note that the recently discovered Q motif (Tanner et al. 2003) is present only in *T. brucei*

A search for DEAD-box protein genes in *T. brucei* identified a single-copy gene encoding a 61-kDa (547 amino acid) polypeptide with a pI of 8.3 (Missel et al. 1997). Compared to other sequences of the DEAD-box family, the trypanosomal protein shares 29% identity with the initiation factor eIF4A of *Saccharomyces cerevisiae* (Foreman et al. 1991), and 31% identity with a DEAD-box helicase of the hyperthermophilic organism *Methanococcus jannaschii* (Story et al. 2001). Among the Kinetoplastida, the identity is even higher. The orthologue proteins share 60% identity between *T. brucei* and *Leishmania major*, and 68% identity between *T. brucei* and *Trypanosoma cruzi* (Fig. 2B). The paralogous genes are localized on non-syntenic chromosomes: chromosome 28 in *L. major* (LmjF28.2080), and chromosome 11 in *T. brucei* (Tb11.01.0610). A chromosome assignment for *T. cruzi* is still pending. The orthologues in *Leishmania* and *T. cruzi* have similar molecular masses with pIs of 8.7 and 7.9, respectively. mHel61p has a putative mitochondrial import sequence, and was experimentally verified as a mitochondrial matrix protein (Missel et al. 1997).

3   mHel61p – Functional Features

T. brucei mHel61p was shown to be expressed as a 2.3-kb transcript in both the insect and the bloodstream lifecycle stages of the parasite. Single-allele, insect-stage knockout trypanosomes behaved indistinguishably from wild-type parasites, and *mhel61* null trypanosomes were found to be viable. However, they multiplied with a significantly reduced growth rate (Missel et al. 1997). This indicated that *mhel61* is not essential in African trypanosomes, although the absence of the protein clearly caused a metabolic deficiency.

On a molecular level, *mhel61* null mutants were characterized (at steady-state conditions) by strongly reduced amounts of edited mRNAs, while never edited mitochondrial mRNAs and nuclear transcripts were unaffected. This phenotype was completely rescued by ectopically re-expressing mHel61p in the knockout parasite cell line (Missel et al. 1997), and therefore demonstrated that mHel61p specifically influences edited transcripts. This suggests a specific involvement of mHel61p in the control and/or stabilization of edited mRNAs. However, mHel61p does not seem to be an integral component of the 20S RNA editing in vitro activity complex (Missel et al. 1997). Most of the polypeptide was identified as non-complexed mitochondrial protein, and only small amounts of mHel61p were found in fractions with an apparent S-value of 20S or higher. Among several possible scenarios, this suggests that mHel61p interacts only transiently, during a specific step of the reaction cycle with the editing machinery.

Support for this interpretation was gained from the result that the RNA editing in vitro activity was not impaired in the mHel61p null mutant. No difference, neither qualitatively nor quantitatively, was observed with respect to the formation of the correct editing product. This implies that the editing machinery can be properly assembled in *mhel61* null trypanosomes, and that it functions indistinguishably from the apparatus in wild-type mitochondria. The result also excludes a potential chaperone function to adjust the secondary structures of gRNA and pre-mRNA molecules prior, or during the editing reaction. Thus, the involvement of mHel61p in the editing process is indirect, functioning in a post-catalytic event that is not monitored in the "single-round" in vitro editing system.

4   mHel61p Structure (Modeling)

As mentioned above, mHel61p of *T. brucei* shares about 30% identity with other bacterial or eukaryotic DExH/D-box proteins. Although the proteins are involved in various biochemical processes, such as RNA degradation, ribosome biogenesis, and RNA splicing, all members of the protein family share nine conserved motifs, including domains for ATP binding, ATP hydrolysis, and RNA binding (Fig. 2). Despite their universal role in so many processes, known high-resolution structures remain limited to only a few DExH/D-box proteins: yeast eIF4A, a prototype minimal DExH/D-box protein (Johnson and McKay 1999; Benz et al. 1999; Caruthers et al. 2000); mjDEAD from *Methanococcus janaschii* (Story et al. 2001); Dhh1p from *Saccharomyces cerevisiae* (Cheng et al. 2005); and the N-terminal domain of BstDEAD from *Bacillus stearothermophilus* (Carmel and Matthews 2004).

Using the structure coordinates of two of the crystallized proteins (eIF4A – PDB file 1fuuB, and mjDEAD – PDB file 1hv8B), a three-dimensional model of mHel61p was calculated by homology modeling (Brecht et al., unpublished data; Fig. 3A). The structure was generated using Geno3D (Combet et al. 2002), utilizing both template files simultaneously. The resulting mHel61p topology is characterized by two α–β domains, each with a RecA-like fold (Ye et al. 2004). The amino-terminal domain carries the DEAD-box motif, while the carboxy-terminal domain contains the helicase motif. In each domain, a core of parallel β-sheets is enclosed by a barrel of α-helices. The model has a "dumbbell" overall structure, with a short linker connecting the two separated domains. The amino-terminal domain is composed of a parallel 5-stranded β-sheet flanked by five helices on one side, and by four helices on the other. The carboxy-terminal domain is also a parallel α–β structure, but with four helices on one side, and three helices on the other, surrounding a central 6-stranded β-sheet. Consistent with their basic RecA-like topology, the N- and C-terminal domains are similar to each other, and superposition of the entire mHel61p structure to eIF4a gave a root mean square deviation (r.m.s.d.) of roughly 3Å. A surface charge calculation revealed that the positively charged amino acids of mHel61p are clustered within the C-terminal domain, which carries the helicase activity, and thus might act as the nucleic acid binding/unwinding domain (Fig. 3B; Schwede et al. 2003; Kopp and Schwede 2004).

5 mHel61p Interacting Proteins

At steady-state conditions, the majority of mHel61p in *T. brucei* mitochondria was found as non-complexed protein, or within protein assemblies of low molecular mass (Missel et al. 1997). However, a subfraction of the protein was identified in association with large mitochondrial RNP complexes, including editosomes (Missel et al. 1997). Furthermore, mHel61p co-purified with active editosome complexes that had been enriched from mitochondrial detergent extracts by biochemical means (Stuart et al. 2002, 2005).

In order to identify direct and indirect interaction partners of mHel61p within these complexes, tandem affinity purification (TAP) experiments were performed (Böhm et al., unpublished data). Starting material was a trypanosome cell line that contained an additional C-terminally TAP-tagged allele of *mhel61* (*mhel61-TAP*).

(A)

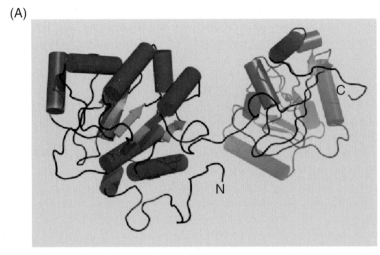

(B)

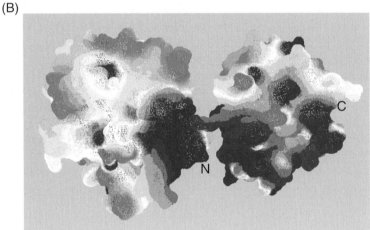

Fig. 3 A Three-dimensional homology model of mHel61p. The structure was generated based on the template PDB data files of eIF4A from *S. cerevisiae* and mjDEAD from *M. jannaschii*. The molecule has a "dumbbell"-like topology: a short linker separates an N- and a C-terminal domain of similar overall structure. A central sheet of parallel β-strands that is caged by a barrel of α-helices characterizes the two domains. **B** Swiss-Pdb Viewer-based representation (Guex and Peitsch 1997) of the electrostatic surface potential of mHel61p. *Red* Negatively charged amino acids, *blue* positively charged amino acids, *white* neutral amino acids

The tagged gene was integrated into one of the transcriptionally silent rDNA-spacer loci of the *T. brucei* genome, and was expressed from a tetracycline inducible PARP (procyclic acidic repetitive protein) promoter. mHel61p interacting proteins were isolated from mitochondrial extracts of insect-stage trypanosomes at near-native conditions by two consecutive affinity chromatography steps (Rigaut et al. 1999). A gel-electrophoretic analysis identified about 20 individual proteins ranging in size between 20–100 kDa. These were identified by mass spectroscopy as almost exclusively editosomal components: the RNA editing endo/exonucleases TbMP90, TbMP61 and TbMP42 (Brecht et al. 2005; Carnes et al. 2005; Trotter et al. 2005), the RNA editing ligase TbMP48 (McManus et al. 2001; Rusché et al. 2001; Gao and Simpson 2003), TbMP46 (Babbarwal et al. 2007), and several other editosomal components of so far unknown function: TbMP99, TbMP67, TbMP63, TbMP64, and TbMP18 (Stuart et al. 2005). Furthermore, the RNA-binding proteins REAP1 (Madison-Antenucci et al. 1998; Madison-Antenucci and Hajduk 2001), RBP38 (Sbicego et al. 2003), and RBP16 (Hayman and Read 1999; Pelletier et al. 2000; Miller and Read 2003; Pelletier and Read 2003) were identified. Lastly, the RNA editing ligase TbMP52 was detected using a self-adenylation assay (Sabatini and Hajduk 1995). Thus, of the 20 editosomal proteins, 15 could be co-purified using a TAP-tagged variant of mHel61p.

However, the affinity-purified complexes were not editing competent. Neither a U-insertion activity nor a U-deletion activity was identified (Böhm et al., unpublished data). As above, this suggests that the mHel61p-containing complexes represent protein assemblies at a specific stage of the editing reaction cycle – likely, after the main catalytic events have taken place.

Lastly, the protein analysis of the mHel61-TAP-containing complexes was suggestive of one additional fact. Almost half of the identified polypeptides still contained their mitochondrial import sequence. This indicates that the removal of the signal sequence is not a prerequisite for the polypeptides to become assembled into mHel61-containing complexes. The phenomenon might reflect that the assembly kinetics of the complexes is significantly faster than the removal of the mitochondrial import sequence from the individual proteins.

6   RNA Unwinding or Editosome Remodeling?

Based on the above-described molecular characterization of the *mhel61* null trypanosome strain, and the structural features of the protein, it is tempting to speculate that mHel61p acts as an RNA helicase for the unwinding of the fully base-paired gRNA/mRNA duplex regions, which presumably form after the completion of all editing events specified by one gRNA (Missel et al. 1997). Failure to disrupt these helical domains could trigger the degradation of the paired RNA structures, and as a consequence, result in reduced amounts of edited mRNAs, a phenotype that is characteristic of the *mhel61*-minus cell line. This would suggest a synchronization of the editing reaction with processes controlling mRNA stability or RNA turnover.

However, based on the fact that mHel61p does not contribute to the "overall" mitochondrial RNA helicase activity in trypanosome mitochondria (Missel et al. 1997), other explanations must be considered. The dsRNA unwinding activity of mitochondrial extracts from mHel61p-minus cells was not different from that of lysates of wild-type mitochondria. Several explanations can account for this observation. For one, other RNA helicases might be present within the *T. brucei* mitochondrion, which could mask the absence of activity in the *mHel61* null mutant. This explanation includes the possibility that more than one helicase participates in the RNA editing process, and candidate proteins can be identified using bioinformatic tools. Similar scenarios have been described in other biochemical processes, most notably in the case of the splicing reaction, which requires the participation of several DEAD-box proteins as auxiliary factors (Schwer and Guthrie 1991; Company et al. 1991). As a consequence, the different mitochondrial RNA helicases might provide, in addition to their specific molecular functions, a basic level of redundant and promiscuous dsRNA unwinding activity (Erickson 1993; Thomas 1993), which in turn might explain the non-lethal phenotype of the *mhel61* null trypanosome strain.

As another explanation, mHel61p might not function as an RNA helicase. Although members of the DExH/D protein family are primarily known for their ability to unwind RNA helices in an ATP-dependent manner, it has become clear in recent years that DExH/D proteins are also involved in the ATP-dependent remodeling of RNA/protein complexes. A large set of experimental data indicates that the structure and composition of ribonucleoprotein complexes is not static, and that changes in RNP structure and/or composition are accurately tuned events to ensure correct RNP function (Staley and Guthrie 1998; Merrick 2004; Nazar 2004). Thus, the undiminished RNA helicase activity in mitochondrial extracts from *mhel61*-minus trypanosomes might indicate that mHel61p is potentially involved in the remodeling of the editosome. This could include processes such as the displacement of editosomal polypeptides from fully base-paired pre-mRNA/gRNA hybrid molecules, or simply a rearrangement of the editing complex to stimulate downstream processes after the U-insertion/deletion reaction has been completed. Unfortunately, our knowledge on how DExH/D proteins remodel RNP complexes is very limited, and to date only a few mechanistic concepts have been put forward (for a review, see Jankowsky and Bowers 2006).

An experimental indication that mHel61p could act as a molecular link to connect the editing reaction to downstream biochemical processes might be hidden in the following fact. The TAP-tagging experiments to isolate mHel61p-containing mitochondrial complexes (cf. above) identified, in addition to the editosomal proteins described above, a polypeptide of the AAA ("ATPases associated with a variety of cellular activities") protease family (Böhm et al., unpublished data). The interaction was further verified in a yeast two-hybrid interaction screen using mHel61p as bait (Müller et al., unpublished data). Members of the AAA family are involved in different cellular functions including the degradation of membrane proteins in bacteria, chloroplasts, and mitochondria (Patel and Latterich 1998). In the membrane, AAA proteases build up large complexes, composed of identical subunits. Their catalytic site faces either the matrix (m-AAA proteases), or the intermembrane space (i-AAA proteases; Leonhard et al. 1996). AAA proteases combine proteolytic and chaperone-like activities, and act as a membrane-integrated quality control system (Langer 2000). The direct interaction of mHel61p with the AAA protease suggests a link between RNA editing and posttranslational processes such as protein degradation.

7   Future Directions

It is clear that the foremost aim must be a complete structural and functional description of the editosomal particle or particles. This will allow a distinction between complexes that are directly involved in the editing reaction, and those that are only intermediates in, for instance, the assembly or disassembly of the active particle. In addition, it will be important to develop improved RNA editing in vitro systems. The assays must be able to monitor multiple rounds of editing in order to analyze more than simply the "core" reaction steps of the process. Within this context, it would be helpful if the editing reaction could be stalled, both in vitro and in vivo, at defined steps of the reaction cycle, to enrich for "functionally" homogenous populations of the different complexes.

In order to specifically address dynamic processes, biochemical and biophysical techniques will have to be applied that can resolve dynamic reaction steps at time scales ranging from several minutes to microseconds, and even less. This will likely involve single-particle detection techniques, and the analysis of dynamic interactions on different levels of complexity:

1. structural rearrangements of RNA and/or protein components as part of the "core" editosome reaction steps, i.e., the exo/endonuclease, TUTase, and ligase reactions;
2. rearrangements on the level of interactions that connect the editing process to other mitochondrial processes such as protein biosynthesis, editosome disassembly, and RNA and protein degradation reactions;
3. structural alterations of the involved molecular components that are triggered upon their interactions with other molecules by, for instance, low molecular mass metabolites or other small regulatory compounds.

From a biophysical point of view, it will be interesting to see whether or not the editosome is a self-assembly system, and if so, what type of RNA–protein and protein–protein interactions have evolved in the pathway(s) towards particle formation. Answers to these types of questions might eventually place the editosome in the company of the two renowned RNP "machines": the ribosome and spliceosome.

Acknowledgements The research described in this chapter was supported by the Howard Hughes Medical Institute (HHMI), the German Research Foundation (DFG), and the Dr. Illing Foundation. H.U.G. is an International Research Scholar of the HHMI. The authors are grateful to E. Schlüter for technical contributions, and to U.F. Müller for the 2H screen.

References

Allen TE, Heidmann S, Reed R, Myler PJ, Göringer HU, Stuart KD (1998) Association of guide RNA binding protein gBP21 with active RNA editing complexes in *Trypanosoma brucei*. Mol Cell Biol 18:6014–6022

176   H.U. Göringer et al.

elAphasizhev R, Simpson L (2001) Isolation and characterization of a U-specific 3′-5′-exonuclease from mitochondria of *Leishmania tarentolae*. J Biol Chem 276:21280–21284

Aphasizhev R, Sbicego S, Peris M, Jang SH, Aphasizheva I, Simpson AM, Rivlin A, Simpson L (2002) Trypanosome mitochondrial 3′ terminal uridylyl transferase (TUTase): the key enzyme in U-insertion/deletion RNA editing. Cell 108:637–648

Aphasizhev R, Aphasizheva I, Nelson RE, Gao G, Simpson AM, Kang X, Falick AM, Sbicego S, Simpson L (2003) Isolation of a U-insertion/deletion editing complex from *Leishmania tarentolae* mitochondria. EMBO J 22:913–924

Babbarwal VK, Fleck M, Ernst NL, Schnaufer A, Stuart K (2007) An essential role of KREPB4 in RNA editing and structural integrity of the editosome in *Trypanosoma brucei*. RNA 13:737–744

Benne R, Van Den Burg J, Brakenhoff JP, Sloof P, Van Boom JH, Tromp MC (1986) Major transcript of the frameshifted *coxII* gene from trypanosome mitochondria contains four nucleotides that are not encoded in the DNA. Cell 46:819–826

Benz J, Trachsel H, Baumann U (1999) Crystal structure of the ATPase domain of translation initiation factor 4A from *Saccharomyces cerevisiae* – the prototype of the DEAD box protein family. Structure 15:671–679

Blom D, Burg Jv, Breek CK, Speijer D, Muijsers AO, Benne R (2001) Cloning and characterization of two guide RNA-binding proteins from mitochondria of *Crithidia fasciculata*: gBP27, a novel protein, and gBP29, the orthologue of *Trypanosoma brucei* gBP21. Nucleic Acids Res 29:2950–2962

Blum B, Bakalara N, Simpson L (1990) A model for RNA editing in kinetoplastid mitochondria: "guide" RNA molecules transcribed from maxicircle DNA provide the edited information. Cell 60:189–198

Brecht M, Niemann M, Schlüter E, Müller UF, Stuart K, Göringer HU (2005) TbMP42, a protein component of the RNA editing complex in African trypanosomes has endo-exoribonuclease activity. Mol Cell 17:621–630

Carmel AB, Matthews BW (2004) Crystal structure of the BstDEAD N-terminal domain: a novel DEAD protein from *Bacillus stearothermophilus*. RNA 10:66–74

Carnes J, Trotter JR, Ernst NL, Steinberg A, Stuart K (2005) An essential RNase III insertion editing endonuclease in *Trypanosoma brucei*. Proc Natl Acad Sci USA 102:16614–16619

Caruthers JM, Johnson ER, McKay DB (2000) Crystal structure of yeast initiation factor 4A, a DEAD-box RNA helicase. Proc Natl Acad Sci USA 97:13080–13085

Cheng Z, Coller J, Parker R, Song H (2005) Crystal structure and functional analysis of DEAD-box protein Dhh1p. RNA 11:1258–1270

Combet C, Jambon M, Deleage G, Geourjon C (2002) Geno3D: automatic comparative molecular modelling of protein. Bioinformatics 18:213–214

Company M, Arenas J, Abelson J (1991) Requirement of the RNA helicase-like protein PRP22 for release of messenger RNA from spliceosomes. Nature 349:487–493

Corell RA, Read LK, Riley GR, Nellissery JK, Allen TE, Kable ML, Wachal MD, Seiwert SD, Myler PJ, Stuart KD (1996) Complexes from *Trypanosoma brucei* that exhibit deletion editing and other editing-associated properties. Mol Cell Biol 16:1410–1418

Erickson HP (1993) Gene knockouts of c-*src*, transforming growth factor β1, and tenascin suggest superfluous, nonfunctional expression of proteins. J Cell Biol 120:1079–1081

Foreman PK, Davis RW, Sachs AB (1991) The *Saccharomyces cerevisiae* RPB4 gene is tightly linked to the TIF2 gene. Nucleic Acids Res 19:2781

Frech GC, Simpson L (1996) Uridine insertion into preedited mRNA by a mitochondrial extract from *Leishmania tarentolae*: stereochemical evidence for the enzyme cascade model. Mol Cell Biol 16:4584–4589

Fuller-Pace FV (1994) RNA helicases: modulators of RNA structure. Trends Cell Biol 4:271–274

Gao G, Simpson L (2003) Is the *Trypanosoma brucei* REL1 RNA ligase specific for U-deletion RNA editing and is the REL2 RNA ligase specific for U-insertion editing? J Biol Chem 278:27570–27574

Guex N, Peitsch MC (1997) SWISS-MODEL and the Swiss-PdbViewer: an environment for comparative protein modeling. Electrophoresis 18:2714–2723

Hans J, Hajduk Sl, Madison-Antenucci S (2007) RNA-editing-associated protein 1 null mutant reveals link to mitochondrial RNA stability. RNA (in press) DOI 10.1261/rna.486107

Hayman ML, Read LK (1999) *Trypanosoma brucei* RBP16 is a mitochondrial Y-box family protein with guide RNA binding activity. J Biol Chem 274:12067–12074

Jankowsky E, Bowers E (2006) Remodeling of ribonucleoprotein complexes with DExD/H-box helicases. Nucleic Acids Res 34:4181–4188

Johnson ER, McKay DB (1999) Crystallographic structure of the amino terminal domain of yeast initiation factor 4A, a representative DEAD-box RNA helicase. RNA 5:1526–1534

Kable ML, Seiwert SD, Heidmann S, Stuart K (1996) RNA editing: a mechanism for gRNA-specified uridylate insertion into precursor mRNA. Science 273:1189–1195

Köller J, Müller UF, Schmid B, Missel A, Kruft V, Stuart K, Göringer HU (1997) *Trypanosoma brucei* gBP21: an arginine-rich mitochondrial protein that binds to guide RNA with high affinity. J Biol Chem 272:3749–3757

Kopp J, Schwede T (2004) The SWISS-MODEL repository of annotated three-dimensional protein structure homology models. Nucleic Acids Res 32:D230–D234

Langer T (2000) AAA proteases: cellular machines for degrading membrane proteins. Trends Biochem Sci 25:247–251

Leonhard K, Herrmann JM, Stuart RA, Mannhaupt G, Neupert W, Langer T (1996) AAA proteases with catalytic sites on opposite membrane surfaces comprise a proteolytic system for the ATP-dependent degradation of inner membrane proteins in mitochondria. EMBO J 15:4218–4229

Linder P (2006) DEAD-box proteins: a family affair – active and passive players in RNP-remodeling. Nucleic Acids Res 34:4168–4180

Madison-Antenucci S, Hajduk SL (2001) RNA editing-associated protein 1 is an RNA binding protein with specificity for preedited mRNA. Mol Cell 7:879–886

Madison-Antenucci S, Sabatini RS, Pollard VW, Hajduk SL (1998) Kinetoplastid RNA-editing-associated protein 1 (REAP-1): a novel editing complex protein with repetitive domains. EMBO J 17:6368–6376

Madison-Antenucci S, Grams J, Hajduk SL (2002) Editing machines: the complexities of trypanosome RNA editing. Cell 108:435–438

Maslov DA, Simpson L (1992) The polarity of editing within a multiple gRNA-mediated domain is due to formation of anchors for upstream gRNAs by downstream editing. Cell 70:459–467

McManus MT, Shimamura M, Grams J, Hajduk SL (2001) Identification of candidate mitochondrial RNA editing ligases from *Trypanosoma brucei*. RNA 7:167–175

Merrick WC (2004) Cap-dependent and cap-independent translation in eukaryotic systems. Gene 332:1–11

Miller MM, Read LK (2003) *Trypanosoma brucei*: functions of RBP16 cold shock and RGG domains in macromolecular interactions. Exp Parasitol 105:140–148

Missel A, Göringer HU (1994) *Trypanosoma brucei* mitochondria contain RNA helicase activity. Nucleic Acids Res 22:4050–4056

Missel A, Souza AE, Nörskau G, Göringer HU (1997) Gene disruption of a mitochondrial DEAD box protein in *Trypanosoma brucei* affects edited mRNAs. Mol Cell Biol 17:4895–4903

Müller UF, Göringer HU (2002) Mechanism of the gBP21-mediated RNA/RNA annealing reaction: matchmaking and charge reduction. Nucleic Acids Res 30:447–455

Müller UF, Lambert L, Göringer HU (2001) Annealing of RNA editing substrates facilitated by guide RNA-binding protein gBP21. EMBO J 20:1394–1404

Nazar RN (2004) Ribosomal RNA processing and ribosome biogenesis in eukaryotes. IUBMB Life 56:457–465

Ochsenreiter T, Hajduk SL (2006) Alternative editing of cytochrome c oxidase III mRNA in trypanosome mitochondria generates protein diversity. EMBO Rep 7:1128–1133

Panigrahi AK, Schnaufer A, Carmean N, Igo RP Jr, Gygi SP, Ernst NL, Palazzo SS, Weston DS, Aebersold R, Salavati R, Stuart KD (2001) Four related proteins of the *Trypanosoma brucei* RNA editing complex. Mol Cell Biol 21:6833–6840

Panigrahi AK, Ernst NL, Domingo GJ, Fleck M, Salavati R, Stuart KD (2006) Compositionally and functionally distinct editosomes in *Trypanosoma brucei*. RNA 12:1038–1049

Patel S, Latterich M (1998) The AAA team: related ATPases with diverse functions. Trends Cell Biol 8:65–71

Pelletier M, Read LK (2003) RBP16 is a multifunctional gene regulatory protein involved in editing and stabilization of specific mitochondrial mRNAs in *Trypanosoma brucei*. RNA 9:457–468

Pelletier M, Miller MM, Read LK (2000) RNA-binding properties of the mitochondrial Y-box protein RBP16. Nucleic Acids Res 28:1266–1275

Pollard VW, Harris ME, Hajduk S (1992) Native mRNA editing complexes from *Trypanosoma brucei* mitochondria. EMBO J 11:4429–4438

Rigaut G, Shevchenko A, Rutz B, Wilm M, Mann M, Séraphin B (1999) A generic protein purification method for protein complex characterization and proteome exploration. Nat Biotechnol 17:1030–1032

Rusché LN, Cruz-Reyes J, Piller KJ, Sollner-Webb B (1997) Purification of a functional enzymatic editing complex from *Trypanosoma brucei* mitochondria. EMBO J 16:4069–4081

Rusché LN, Huang CE, Piller KJ, Hemann M, Wirtz E, Sollner-Webb B (2001) The two RNA ligases of the *Trypanosoma brucei* RNA editing complex: cloning the essential band IV gene and identifying the band V gene. Mol Cell Biol 21:979–989

Sabatini R, Hajduk SL (1995) RNA ligase and its involvement in guide RNA/mRNA chimera formation. Evidence for a cleavage-ligation mechanism of *Trypanosoma brucei* mRNA editing. J Biol Chem 270:7233–7240

Sbicego S, Alfonzo JD, Estevez AM, Rubio MA, Kang X, Turck CW, Peris M, Simpson L (2003) RBP38, a novel RNA-binding protein from trypanosomatid mitochondria, modulates RNA stability. Eukaryot Cell 2:560–568

Schmid SR, Linder P (1992) D-E-A-D protein family of putative RNA helicases. Mol Microbiol 6:283–291

Schnaufer A, Ernst NL, Palazzo SS, O'Rear J, Salavati R, Stuart K (2003) Separate insertion and deletion subcomplexes of the *Trypanosoma brucei* RNA editing complex. Mol Cell 12:307–319

Schumacher MA, Karamooz E, Zikova A, Trantirek L, Lukes J (2006) Crystal structures of *T. brucei* MRP1/MRP2 guide-RNA binding complex reveal RNA matchmaking mechanism. Cell 126:701–711

Schwede T, Kopp J, Guex N, Peitsch MC (2003) SWISS-MODEL: an automated protein homology-model server. Nucleic Acids Res 31:3381–3385

Schwer B, Guthrie C (1991) PRP16 is an RNA-dependent ATPase that interacts transiently with the spliceosome. Nature 349:494–499

Seiwert SD, Stuart K (1994) RNA editing: transfer of genetic information from gRNA to precursor mRNA in vitro. Science 266:114–117

Seiwert SD, Heidmann S, Stuart K (1996) Direct visualization of uridylate deletion *in vitro* suggests a mechanism for kinetoplastid RNA editing. Cell 84:831–841

Staley JP, Guthrie C (1998) Mechanical devices of the spliceosome: motors, clocks, springs, and things. Cell 92:315–326

Story RM, Li H, Abelson JN (2001) Crystal structure of a DEAD box protein from the hyperthermophile *Methanococcus jannaschii*. Proc Natl Acad Sci USA 13:1465–1470

Stuart K, Panigrahi AK, Schnaufer A, Drozdz M, Clayton C, Salavati R (2002) Composition of the editing complex of *Trypanosoma brucei*. Philos Trans R Soc Lond B Biol Sci 357:71–79

Stuart KD, Schnaufer A, Ernst NL, Panigrahi AK (2005) Complex management: RNA editing in trypanosomes. Trends Biochem Sci 30:97–105

Tanner NK, Cordin O, Banroques J, Doere M, Linder P (2003) The Q motif: a newly identified motif in DEAD box helicases may regulate ATP binding and hydrolysis. Mol Cell 11:127–138

Thomas JH (1993) Thinking about genetic redundancy. Trends Genet 9:395–399

Trotter JR, Ernst NL, Carnes J, Panicucci B, Stuart K (2005) A deletion site editing endonuclease in *Trypanosoma brucei*. Mol Cell 20:403–412

Vanhamme L, Perez-Morga D, Marchal C, Speijer D, Lambert L, Geusken M, Alexandre S, Ismaïli N, Göringer HU, Benne R, Pays E (1998) *Trypanosoma brucei* TBRGG1, a mitochondrial oligo(U)-binding protein that co-localizes with an *in vitro* RNA editing activity. J Biol Chem 273:21825–21833

Ye J, Osborne AR, Groll M, Rapoport TA (2004) RecA-like motor ATPases – lessons from structures. Biochim Biophys Acta 1659:1–18

The Function of RNA Editing in Trypanosomes

Torsten Ochsenreiter and Stephen Hajduk (✉)

> *"Infinite diversity in infinite combinations ... symbolizing the elements that create truth and beauty."*
>
> Commander Spock
> Star Trek.

Abstract RNA editing of mitochondrial mRNAs in trypanosomes is characterized by the post-transcriptional insertion or deletion of uridine residues. This process is directed by a diverse family of small, guide RNAs (gRNAs). The trypanosome mitochondrial genome contains nine genes for mRNAs that are extensively edited, three genes for mRNAs that are edited at a limited number of sites, and six genes that encode mRNAs, which are not edited. Editing is essential in both insect and bloodstream developmental stages of the parasite, and results in the formation of initiation codons and extended open reading frames in mRNA, necessary for maintaining mitochondrial electron transport and oxidative phosphorylation. In essence, RNA editing is responsible in producing functional mRNAs so that conventional mitochondrial proteins can be made. Another potential raison d'être for RNA editing in trypanosomes has recently been demonstrated. Trypanosome mitochondrial mRNAs can be differentially edited, producing mRNAs with extended open reading frames. This suggests that multiple proteins can be produced from a single mitochondrial gene. A product of alternative editing of the cytochrome oxidase subunit III (COIII) mRNA encodes a mitochondrial membrane protein with a unique, arginine-rich N-terminal

Department of Biochemistry and Molecular Biology, University of Georgia, 120 Green Street, Athens, GA 30602-7229, USA; shajduk@bmb.uga.edu

H.U. Göringer (ed.), *RNA Editing. Nucleic Acids and Molecular Biology 20*
© Springer-Verlag Berlin Heidelberg 2008

sequence and the C-terminal sequence of COIII. Alternative mRNA editing thus expands the repertoire of mitochondrial proteins in trypanosomes. This review addresses the functions of RNA editing in both RNA repair and generation of protein diversity in these organisms.

1 Introduction

RNA editing in the kinetoplastid Protozoa is a post-transcriptional process that modifies many mitochondrial pre-mRNAs by the insertion or deletion of uridine nucleotides. The initial description of RNA editing, over 20 years ago by Benne and co-workers (Benne et al. 1986), was greeted with considerable skepticism. In their landmark paper, the addition of a modest four uridines within the pre-mRNA of cytochrome oxidase II was reported. Troubling was the lack of an apparent template for these added nucleotides. Shortly after the discovery of mRNA editing, Stuart and co-workers reported the extensive editing of cyto-chrome oxidase III that resulted in the insertion of 547 uridines and the deletion 41 uridine residues (Feagin et al. 1988). These two papers brought considerable attention to this remarkable biological phenomenon, and raised fundamental questions about the origin, mechanism, and function of RNA editing in trypanosomes.

There have been a number of recent reviews on the mechanism and the origin of RNA editing, and the composition of the editing machinery, the editosome (Stuart and Panigrahi 2002; Madison-Antenucci et al. 2002; Simpson et al. 2003, 2004; Stuart et al. 2005). In this chapter, we address the underlying question – why edit? In particular, we will emphasize some of the unique biological features of trypanosomes, and the role RNA editing plays in molding these traits.

2 Why Kinetoplasts?

The defining feature of kinetoplastids is the unique organization of their mitochondrial DNA, called the kinetoplast. Initially seen by cytologists as a basophilic granule at the base of the flagellum, the kinetoplast is of historical note as the first extra-nuclear DNA described. The unusual structure of the kinetoplast was first suggested by transmission electron microscopy of fixed and sectioned trypanosomes, which revealed a highly organized fibrous structure within double mitochondrial membranes at the base of the flagellum (Fig. 1, top part). The development of methods to isolate and image pure kinetoplast DNA (kDNA) revealed the truly remarkable nature of the structure of this DNA. The kDNA of trypanosomes is composed of thousands of small, circular DNA molecules called minicircles, topologically interlocked to form a huge network structure (Fig. 1, bottom part; Riou and Delain 1969). Later studies also revealed the presence of

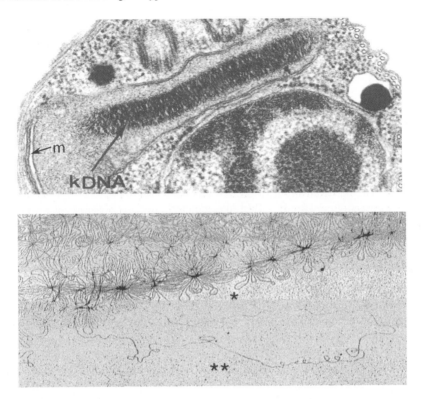

Fig. 1 Kinetoplast DNA of trypanosomes. The mitochondrial DNA of *T. brucei* is a highly organized structure, called the kinetoplast, which is localized to a portion of the mitochondrion adjacent to the flagellum. *Top* Transmission electron microscopy reveals the compact organization of the kinetoplast, and its close proximity to the mitochondrial membrane (*m*). *Bottom* Purified kDNA visualized by transmission electron microscopy shows the catenated network of minicircles (*) and maxicircles (**)

another, larger DNA component within the kDNA network, called maxicircles (Steinert and Assel 1975; Kleisen et al. 1976). Remarkably, the combined minicircle and maxicircle genomes account for 10–20% of the total trypanosome DNA. Other eukaryotes have small mitochondrial genomes representing less than 1% of their total DNA, typically encoding mitochondrial rRNAs, tRNAs, and a few components of the electron transport and ATP synthesis machinery. So, why would trypanosomes, simple eukaryotes, need such a large amount of mitochondrial DNA?

Puzzling also was the organization of the kinetoplast. Not only do trypanosomes relegate a large fraction of their genome to mitochondrial DNA, but it is packaged as a tangled mass of tiny circles and a few larger molecules. The faithful replication and segregation of such a "chromosome" seemed unfathomable. Taken together, these early studies led to the inevitable question – why do trypanosomes have kinetoplasts?

3   Why Trypanosomes?

Over decades of study, trypanosomes have developed as somewhat of a "model" system. These early-branching eukaryotes have offered an important glimpse of the evolution of several important processes. In addition, trypanosomes are also medically important. Subspecies of *T. brucei* are the causative agents of both Nagana, a chronic wasting disease in cattle, and African sleeping sickness in humans. *T. brucei* continues to be a major cause of cattle death throughout equatorial Africa, and human sleeping sickness has re-emerged over the past decade as a serious human disease resulting in over 60,000 deaths per year (WHO 2006).

One unique aspect of trypanosome biology is the developmental regulation of the pathways for energy production. Trypanosomiasis is a vector-borne disease passed by the bite of an infected tsetse fly (*Glossina* spp.) to a mammalian host (WHO 2006). *T. brucei* undergoes a number of biochemical and morphological changes during its development in the mammalian host and the insect vector (Matthews 2005). The most dramatic changes are associated with the single, large mitochondrion of the parasite. During the developmental stages of *T. brucei* in the insect vector, the procyclic form (PF), the parasite has a fully developed mitochondrion containing numerous cristae, electron transport complexes, and ATP synthase. Energy production in the PF trypanosome is largely mitochondrial, and is performed by coupled electron transport and oxidative phosphorylation (Besteiro et al. 2005). When trypanosomes are transmitted to the mammalian host, mitochondrial activities are rapidly repressed, and in the fully differentiated bloodstream form (BF), the mitochondrion is reduced in volume and is acristate. The BF trypanosome derives all of its ATP by substrate-level phosphorylation through glycolysis, which occurs mainly in a specialized organelle, the glycosome (Parsons 2004; Michels et al. 2006). The ability to developmentally regulate mitochondrial function is a unique aspect of trypanosome biology. No other organism has developed regulatory mechanisms to allow the coordinated assembly of mitochondrial activities during a normal life cycle. This unique process requires novel regulators; RNA editing is one such regulator.

4   What Is RNA Editing?

When RNA editing was discovered in trypanosomes by Benne and co-workers (Benne et al. 1986), it was viewed as an interesting and surprising quirk of trypanosomes that, at best, might also be found in the organelles of other "simple" single-celled organisms. The real surprise has been the range of organisms that edit RNA, and the diversity of RNA editing mechanisms and substrates. This diversity has lead to a redefinition of RNA editing to include any RNA modification that changes the sequence of a transcript relative to its gene (Gott and Emeson 2000). Virtually all eukaryotes "edit" RNA; rRNA, mRNAs, microRNAs, and tRNAs have been identified as substrates for editing (Schaub and Keller 2002; Blanc and Davidson 2003; Wedekind et al. 2003; Luciano et al. 2004; Shikanai 2006).

Trypanosome RNA editing requires the coordinated interactions of RNA products from minicircles and maxicircles, and proteins encoded by nuclear genes that are imported into the mitochondria. The trypanosome mitochondrial genome, maxicircles and minicircles, is physically associated within the kDNA network. Maxicircles are homologous to the mitochondrial genome of other eukaryotes, and were the first trypanosome genome completely sequenced (Benne et al. 1983; Eperon et al. 1983; Hensgens et al. 1984; Payne et al. 1985). Deciphering the coding information of this modest 22-kb circular DNA proved exceedingly difficult. Initial sequence analysis revealed several predicted mitochondrial genes, including the mitochondrial ribosomal RNAs (rRNAs), cytochrome oxidase subunit I (COI), and NADH dehydrogenase subunits 1, 4, and 5 (ND1, ND4, and ND5). Further maxicircle sequence analysis revealed partial open reading frames for cytochrome b (CYb) and cytochrome oxidase II (COII). However, the sequence analysis failed to reveal several essential and invariant mitochondrial genes including tRNAs and genes for respiratory complex proteins, cytochrome oxidase III (COIII), NADH dehydrogenase subunits 7, 8, 9, and the ATP synthase subunit 6.

Only after Benne's discovery and further sequencing of mitochondrial cDNAs did it become clear that many of the maxicircle encoded mRNAs are substrates for RNA editing (Hajduk et al. 1993, 1997; Estevez and Simpson 1999). Some mRNAs, such as COII, are modified by a relatively small number of uridine insertions (Benne et al. 1986). Other mRNAs, such as COIII, undergo extensive RNA editing resulting in the insertion and deletion of hundreds of uridines (Feagin et al. 1988).

While the maxicircle component contributes at least some of the coding information anticipated for a mitochondrial genome, the function of the minicircle genome was less obvious. In 1990, nearly 4 years after the discovery of editing of COII mRNA, Blum and Simpson proposed that small mitochondrial RNAs could provide the information needed for the editing of maxicircle encoded pre-mRNAs (Blum et al. 1990). The "guide RNA" hypothesis was born. Although the first gRNAs were discovered on the maxicircle of *Leishamia tarentolae*, it soon became clear that most of the gRNAs are encoded on the minicircles of *T. brucei* and *L. tarentolae*, thus resolving a major question in trypanosome biology (Sturm and Simpson 1990; Pollard et al. 1990; Pollard and Hajduk 1991; van der Spek et al. 1991).

Guide RNAs direct the editing machinery to the substrate pre-mRNA, and provide a template for the correct insertion and deletion of uridines in pre-mRNAs. Guide RNAs range in size from 30–50 nucleotides, and are encoded mainly by minicircles (Hong and Simpson 2003; Ochsenreiter et al. 2007; Fig. 2). The 5′ terminal 4–12 nts of each gRNA can base pair with a pre-edited mRNA substrate, forming a perfect "anchor" immediately adjacent to an RNA editing site. The formation of this RNA duplex between the gRNA and pre-mRNA directs the formation of the editing machinery, and thus serves as the recognition element for editing and is critical to fidelity. Each gRNA directs the successive insertion or deletion of uridines over a region of about 30 nts, and results in the formation of edited mRNA. The fully edited mRNA is perfectly complementary with its cognate gRNAs.

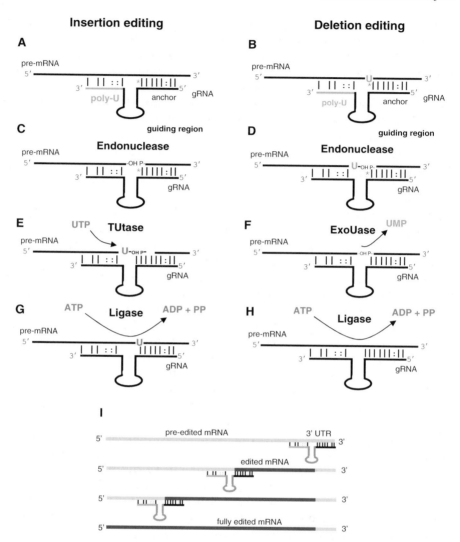

Fig. 2 General mechanism for insertion and deletion mRNA editing in trypanosomes. **A, B** Guide RNAs bind to their cognate pre-mRNAs via the 8–12 nucleotide anchor region on the 5′ end of the gRNA molecule. Binding is facilitated mainly through Watson Crick base pairing (*solid lines*) and G:U base pairings (:). The 3′ poly U tail of the gRNA stabilizes the gRNA: mRNA interaction. **C, D** The first mismatch (*) between gRNA and pre-mRNA, 3′ of the anchor duplex, is the signal for the endonuclease to cleave the pre-mRNA, resulting in a 3′ and 5′ mRNA cleavage product. **E** Post-cleavage insertional editing adds one uridine to the 3′ hydroxyl of the 5′ cleavage product by a terminal uridylyl transferase (TUTase). **F** In deletion editing, the endonuclease step is followed by an exonuclease that removes one uridine residue from the 3′ end of the 5′ cleavage product. **G, H** After insertion or deletion, the 5′ and 3′ cleavage products are joined by an RNA ligase using ATP hydrolysis. **I** Insertion/deletion editing progresses 3′ to 5′ on the pre-mRNA. The first gRNA usually anchors in the non-edited 3′ UTR of the pre-mRNA

Despite the small size of the individual minicircles, the "complete genome" is very large. Current estimates put the complexity of minicircles at greater than 500 sequence classes per kDNA network (Ochsenreiter et al. 2007). Since each minicircle contains on average three gRNA genes, a single trypanosome has at least 1,500 gRNA genes. The significance of this extreme sequence heterogeneity has recently been shown for alternative editing of mRNA that contributes to mitochondrial protein diversity in these organisms (Ochsenreiter and Hajduk 2006).

5   How to Edit RNA?

Mitochondrial mRNA editing in trypanosomes is an enzymatic cascade catalyzed by large protein complexes that sediment at approximately 20S on glycerol gradients, and are termed editosomes (Pollard et al. 1992; Panigrahi et al. 2006; reviewed in Simpson et al. 2003; Stuart et al. 2005). In addition, larger protein complexes, also containing gRNAs and mRNAs, sediment between 35–40S and may be editosomes actively engaged in mRNA editing (Pollard et al. 1992; reviewed in Madison-Antenucci et al. 2002). The protein composition of catalytically active 20S editosomes has been extensively investigated, mainly by the Stuart group (Madison-Antenucci et al. 2002; Ernst et al. 2003; Panigrahi et al. 2003a, b; Schnaufer et al. 2003), and over 20 nuclear encoded mitochondrial proteins have been identified, and their roles in RNA editing systematically evaluated by RNA interference and gene knockout (Schnaufer et al. 2001; Drozdz et al. 2002; Brecht et al. 2005; Salavati et al. 2006). The enzymatic activities necessary for both uridine nucleotide insertion and deletion editing include editing site-specific endonuclease(s), terminal uridylyl transferase(s) or uridine-specific exoribonuclease, and RNA ligase(s). The coordinated activity of these enzymes is facilitated by assembly into the editosome. The mechanism of editosome assembly and editing site selection remains unresolved, but the prevailing theory is that the 20S editosomes are heterogeneous, and editosomes responsible for insertional or deletional editing are distinct species (Panigrahi et al. 2006).

Editing of an mRNA is initiated by the formation of a short anchor duplex between 5′ regions of a gRNAs and a region immediately 3′ to the editing site within an mRNA (Blum and Simpson 1990; Blum et al. 1990; Pollard et al. 1990; Fig. 2). The unpaired sequence of the pre-mRNA, immediately adjacent to the anchor duplex, is recognized by the 20S editosome and is cleaved by either an addition or deletion site-specific endoribonuclease (Cruz-Reyes and Sollner-Webb 1996; Kable et al. 1996). Endonuclease cleavage of the pre-mRNA generates a 5′ phosphate and 3′ hydroxyl. This cleaved pre-mRNA is the substrate for the second enzyme of the editing cascade, and depending on the information provided by the gRNA, uridines can either be deleted or added. Each round of editing culminates with the joining of the cleaved pre-mRNA, and extension of the base pairing between the gRNA and pre-mRNA (Fig. 2).

Base pairing between the gRNA and cognate mRNA plays a critical role in the recognition of editing sites on the pre-mRNA, and also in specifying the correct number of uridines to be added or deleted. Several gRNAs are needed to edit many pre-mRNAs, and in extensively edited mRNA the first editing site is often near the 3' untranslated region of the mRNA (Cruz-Reyes and Sollner-Webb 1996; Kable et al. 1996; Hong and Simpson 2003; Ochsenreiter et al. 2007).

Many of the basic features of uridine insertion and deletion RNA editing have been defined. Initial questions concerning the template for editing and the mechanism have been resolved with the discovery of gRNAs, editing enzymes, and editosomes. An ever-growing list of editosome proteins and associated proteins promises to provide the components necessary to unravel the detailed catalytic mechanism of RNA editing (reviewed in Stuart et al. 2005). Atomic structures are now available for two of the editing proteins and one gRNA binding protein (Deng et al. 2004, 2005; Schumacher et al. 2006; Stagno et al. 2007). Structural analyses of other editing proteins and the 20S editosome are underway.

6 Why RNA Editing?

The type of uridine insertion and deletion RNA editing in trypanosomes is not found outside of the order Kinetoplastidae. Therefore, it is unlikely that this form of RNA editing evolved in an early ancestral organism, but more likely was an isolated event that occurred following the divergence of the kinetoplastid lineage (Gray 1994). At first glance, RNA editing appears to be overly complex, inefficient, and altogether unnecessary; yet, RNA editing is essential to trypanosomes.

Many of the maxicircle genes lack complete open reading frames necessary for the production of mitochondrial proteins needed for electron transport and oxidative phosphorylation. RNA editing provides the missing sequence information by forming initiation codons, correcting frameshift mutations, and creating entire open reading frames (Fig. 3). Thus, a clear function for RNA editing in trypanosome mitochondria is to fill in missing information, lacking in maxicircle genes, by uridine insertion or deletion in mRNAs. Furthermore, the proteomic analysis of the 20S editosome showed that a number of editosome proteins had highest sequence similarity to DNA repair enzymes (Panigrahi et al. 2003a, b). Together, these observations lead to the conclusion that an RNA repair function served as the evolutionary raison d'être for RNA editing (Gray 2003). While we agree that a basic function of uridine insertion and deletion RNA editing is to create start and stop condons, correct for frameshifts, and create open reading frames for some of the conventional mitochondrial genes, we believe it is an oversimplification to consider this strictly a repair mechanism.

Other organisms also use RNA editing to change gene sequences at an RNA level; however, the generation of protein diversity, rather than "repair", seems the primary function. In mammals, adenosine to inosine (A to I) editing allows for the formation of multiple RNA and protein products from a single genomic locus.

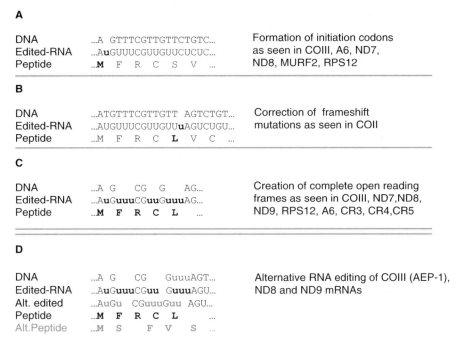

Fig. 3 Function of trypanosome mRNA editing. **A** The formation of initiation codons (and also stop codons). **B** The correction of frameshift mutations, and **C** the creation of entire open reading frames. **D** These three basic functions can lead either to the creation of a bona fide mRNA or to an alternative product (*bold u* inserted uridine residues, *capital bold letters* the corresponding amino acid sequence, *underlined letters* alternative mRNA editing). The ability to create numerous open reading frames from a pre-edited mRNA is a feature of trypanosome RNA editing. This allows the parasite to expand the coding potential of the mitochondrial genome

This expands the diversity of protein products and biological function (Knight and Bass 2002). For example, A to I editing in mammals can give rise to subtle changes in neuronal receptors and ion channels that lead to different protein isoforms, with profound physiological and pathophysiological changes (Sommer et al. 1991). Interestingly, these changes often have only limited region-specific effects on protein structure, but dramatic effects on functional properties (Singh and Emeson 2001). Mammals also developmentally regulate mRNA editing, the best-studied example being the C to U editing of the apolipoprotein B mRNA in the intestine and liver, which gives rise to functionally different isomers of the protein (Chen et al. 1987).

Mitochondrial mRNA editing in trypanosomes also expands the sequence diversity of mRNAs from a single gene (Koslowsky et al. 1990; Souza et al. 1992; Corell et al. 1994; Read et al. 1994; Ochsenreiter and Hajduk 2006; Table 1). The first description of alternatively edited mRNAs in trypanosomes came shortly after the discovery of extensive editing of trypanosome mRNA by the Stuart laboratory (Feagin et al. 1988). Koslowsky and co-workers showed that NADH dehydrogenase

Table 1 Mitochondrial mRNA editing in trypanosomes

Maxicircle gene[a]	Editing[b]	Developmental regulation[c]	Alternative transcripts[d]	Alternative open reading frames (ORFs)[e]	Initiation codon[f]	Reference
Complex I						
ND1	Not edited	Unknown	Not known		UUG	
ND3	Insertion: 210; deletion: 13	BF and PF	IPF and BS 3′ region alt. ed.	Multiple C-termini	UUG	Read et al. (1994)
ND4	Not edited	BF-abundant			AUG	
ND5	Not edited	BF-abundant			AUG	
ND7	Insertion: 553; deletion: 89	BF-abundant; BF-& PF-5′ region; BF-3′ region	PF-5′ region edited		AUG	Koslowsky et al. (1990)
ND8	Insertion: 259; deletion: 46	BF-abundant	BF-5′ region pre-edited	Two alternative ORFs	AUG (alt. ORF1 UUG, ORF2 GUG)	Souza et al. (1992); Ochsenreiter and Hajduk (unpubl. data)
ND9	Insertion: 345; deletion: 20	BF-abundant	BF-5′ region alt. ed.	One alternative ORF	AUG (alt. ORF GUG or UUG)	Souza et al. (1993); Ochsenreiter and Hajduk (unpubl. data)
Complex III						
CYb	Insertion: 39; deletion: none	PF-abundant			AUU or UUG	Feagin et al. (1987)
Complex IV						
COI	Not edited	PF-abundant			AUG	
COII	Insertion: 4; deletion: none	PF-abundant	Not known		AUG	Benne et al. (1986)
COIII	Insertion: 547; deletion: 41	BF and PF	AEP-1	Three alternative ORFs	AUG	Feagin et al. (1988); Ochsenreiter and Hajduk (2006)

	Editing[b]	Expression[c]	Alt. editing[d]	ORFs[e]	Initiation codon[f]	Reference
Complex V						
A6	Insertion: 447; deletion: 28	BF and PF	Not known		AUG	Bhat et al. (1990)
Ribosome						
S12	Insertion: 132; deletion: 28	BF-abundant	Not known		AUG	Marchal et al. (1993)
Unknown function						
MURF1	Not edited				UUG	
MURF2	Insertion: 26; deletion: 4	BF-abundant	Not known		AUG	Feagin and Stuart (1988)
MURF5	Not edited				AUG	
CR3	Insertion: 148; deletion: 13	BF-abundant	Not known	Two alternative ORFs	AUA or UUG	Stuart et al. (1997)
CR4	Insertion: 325; deletion: 40	BF-abundant	In PF, the 5' region alt. ed.		AUG (alt. ORF UUG)	Corell et al. (1994)
rRNAs						
12S rRNA	3'-Poly U tail	BF and PF				Adler et al. (1991)
9S rRNA	3'-Poly U tail	BF and PF				Adler et al. (1991)

[a] Abbreviations: ND, NADH ubiquinone oxidoreductase subunits 1, 3, 4, 5, 7, 8, 9; CYb, apocytochrome b; CO, cytochrome oxidase subunits I, II, and III; A6, ATP synthase subunit 6; S12, ribosomal protein S12; MURF, maxicircle unidentified reading frame; CR, G- versus C-strand biased gene subunits 3 and 4; 12S and 9S, mitochondrial ribosomal RNAs

[b] Number of uridines inserted or deleted for each mRNA

[c] BF, bloodstream form, and PF, insect (procyclic) forms of T. brucei

[d] Alternatively edited (alt. ed.) mRNAs that differ from the predicted bona fide mRNA

[e] Alternatively edited mRNAs can have unique open reading frames (ORFs)

[f] Canonical and alternative initiation codons that result in the longest ORF are shown

subunit 7 (ND7) mRNA was extensively edited in both BF and PF trypanosomes, and that editing was differentially regulated in the two developmental stages of the parasite (Koslowsky et al. 1990). The ND7 mRNA contains two distinct domains, the 5' domain being edited in both BF and PF trypanosomes, while the 3' domain was edited only in the BF trypanosomes. Interestingly, editing of the 5' domain and not the 3' domain, the situation seen in PF trypanosomes, results in an inframe termination codon (UAG) and the formation of a predicted truncated ND7 isoform. Other studies have reported the formation of alternatively edited mRNAs, but the ability of these mRNAs to be translated to alternate mitochondrial proteins was not shown until only recently (Ochsenreiter and Hajduk 2006).

Cytochome c oxidase is the mitochondrial terminal oxidase complex in most eukaryotes, including the PF of African trypanosomes. In the BF trypanosomes, mitochondrial electron transport is greatly reduced and oxidative phosphorylation is absent. The BF trypanosomes express a plant-like, alternative terminal oxidase (TAO) that localizes to the mitochondrion where its primary function is to facilitate the re-oxidation of NADH generated during glycolysis. The metabolic changes associated with the parasite's life cycle are reflected in changes in the abundance of both mitochondrial and nuclear encoded subunits of cytochrome c oxidase – with one notable exception. Cytochrome c oxidase subunit III mRNA is abundant in BF trypanosomes, and is also extensively edited (Abraham et al. 1988). Since the only known function for COIII is oxidative phosphorylation, and the mRNAs for all other subunits of the oxidase are missing in BF trypanosomes, this is surprising. Sequence analysis of COIII cDNAs from BF trypanosomes revealed that the sequences were highly diverse and also contained partially edited mRNAs with extended open reading frames (Ochsenreiter and Hajduk 2006). One of these COIII mRNAs contained a long open reading frame including both pre-edited and edited sequences. A gRNA was identified that could direct alternative editing at the junction between the edited and pre-edited sequences. The predicted protein encoded by this alternatively edited COIII mRNA is composted of a unique hydrophilic amino-terminal domain and five membrane-spanning domains of COIII in the carboxyl terminus (Fig. 4). The existence of a protein product from the alternatively edited COIII mRNA was established with antibodies raised against synthetic peptides based on the predicted sequence from the alternatively edited mRNA. An alternatively edited protein (AEP-1) was detected in cell fractionation studies as a mitochondrial membrane protein, and in situ by immunofluorescence microscopy (Ochsenreiter and Hajduk 2006). While the alternative editing of COIII mRNA in BF trypanosomes was clearly established in these studies, the biological function of AEP-1 remains to be elucidated.

How prevalent is alternative editing of trypanosome mitochondrial mRNA? Based on the extensive sequence analysis of full-length COIII mRNAs, it appears that there is a disproportionate number of alternatively edited mRNAs in BF trypanosomes relative to fully edited COIII mRNA (Ochsenreiter and Hajduk, unpublished data). While it is possible that the fate of most differentially edited mRNAs is degradation, the results with AEP-1 suggest that at least some of these transcripts are stable, translatable mRNAs. These studies indicate that RNA editing contributes to a new level of genetic diversity in trypanosome mitochondria.

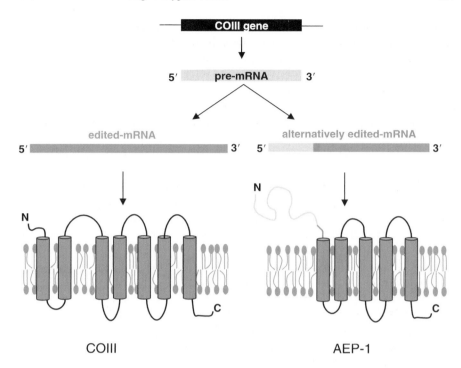

Fig. 4 Alternative RNA editing in trypanosomes. Transcription of COIII gene leads to pre-edited mRNA (*black with white dots*). This mRNA can be edited either to the bona fide COIII sequence (*grey*), or via an alternative pathway to a novel mRNA. The alternatively edited mRNA contains pre-edited (*black with white dots*), alternatively edited (*striped grey*), and bona fide edited sequences (*grey*). An open reading frame is formed that spans the pre-edited, junction and edited region of this transcript. The protein product, AEP-1, from this open reading frame is a mitochondrial membrane in *T. brucei*, indicating that alternative mRNA editing produces alternate, functional proteins

The discovery of alternatively edited mRNAs and AEP-1 suggests a mechanism for increased diversity of trypanosome mitochondrial proteins. The novel proteins generated by RNA editing might be important, since members of the Kinetoplastidae have a number of unique mitochondrial features including the organization of the kDNA network, and its replication and segregation. The unique organization of the mitochondrial DNA of trypanosomes presents a significant challenge for the organism to faithfully replicate thousands of DNA molecules, and equally segregate the molecules to a daughter cell each cell cycle. All kinetoplastids have evolved mechanisms to facilitate the replication and segregation of the kDNA. Replication of kinetoplast minicircles and maxicircles involves an elaborate orchestration of decatenation, replication, and reattachment (reviewed in Klingbeil and Englund 2004). This process is non-random and molecular tags, in the form of gapped circles, ensure that each circle is replicated once and is efficiently reattached (reviewed in Liu et al. 2005). Both the kDNA network and replicating minicircles and maxicircles are physically restricted to the region of the mitochondrion immediately adjacent to the base of the flagellum (reviewed in Liu et al. 2005). The flagellar association, which traverses the double mitochondrial membrane, may be critical to ensure the full complement of minicircles is replicated and reattached to the growing kDNA network. It is likely that the proteins involved in kDNA membrane attachment and segregation will be unique to the kinetoplastids, and may require proteins produced within the mitochondrion. The recent discovery of alternative RNA editing suggests that novel protein sequences could be created that carry out these "kinetoplast-specific" functions.

References

Abraham JM, Feagin JE, Stuart K (1988) Characterization of cytochrome c oxidase III transcripts that are edited only in the 3' region. Cell 55:267–272

Adler BK, Harris ME, Bertrand KI, Hajduk SL (1991) Modification of *Trypanosoma brucei* mitochondrial rRNA by posttranscriptional 3' polyuridine tail formation. Mol Cell Biol 11:5878–5884

Benne R, De Vries BF, Van den Burg J, Klaver B (1983) The nucleotide sequence of a segment of *Trypanosoma brucei* mitochondrial maxi-circle DNA that contains the gene for apocytochrome b and some unusual unassigned reading frames. Nucleic Acids Res 11:6925–6941

Benne R, Van den Burg J, Brakenhoff JP, Sloof P, Van Boom JH, Tromp MC (1986) Major transcript of the frameshifted coxII gene from trypanosome mitochondria contains four nucleotides that are not encoded in the DNA. Cell 46:819–826

Besteiro S, Barrett MP, Riviere L, Bringaud F (2005) Energy generation in insect stages of *Trypanosoma brucei*: metabolism in flux. Trends Parasitol 21:185–191

Bhat GJ, Koslowsky DJ, Feagin JE, Smiley BL, Stuart K (1990) An extensively edited mitochondrial transcript in kinetoplastids encodes a protein homologous to ATPase subunit 6. Cell 61:885–894

Blanc V, Davidson NO (2003) C-to-U RNA editing: mechanisms leading to genetic diversity. J Biol Chem 278:1395–1398

Blum B, Simpson L (1990) Guide RNAs in kinetoplastid mitochondria have a nonencoded 3' oligo(U) tail involved in recognition of the preedited region. Cell 62:391–397

Blum B, Bakalara N, Simpson L (1990) A model for RNA editing in kinetoplastid mitochondria: "guide" RNA molecules transcribed from maxicircle DNA provide the edited information. Cell 60:189–198

Brecht M, Niemann M, Schluter E, Muller UF, Stuart K, Göringer HU (2005) TbMP42, a protein component of the RNA editing complex in African trypanosomes, has endo-exoribonuclease activity. Mol Cell 17:621–630

Chen SH, Habib G, Yang CY, Gu ZW, Lee BR, Weng SA, Silberman SR, Cai SJ, Deslypere JP, Rosseneu M et al. (1987) Apolipoprotein B-48 is the product of a messenger RNA with an organ-specific in-frame stop codon. Science 238:363–366

Corell RA, Myler P, Stuart K (1994) *Trypanosoma brucei* mitochondrial CR4 gene encodes an extensively edited mRNA with completely edited sequence only in bloodstream forms. Mol Biochem Parasitol 64:65–74

Cruz-Reyes J, Sollner-Webb B (1996) Trypanosome U-deletional RNA editing involves guide RNA-directed endonuclease cleavage, terminal U exonuclease, and RNA ligase activities. Proc Natl Acad Sci USA 93:8901–8906

Deng J, Schnaufer A, Salavati R, Stuart KD, Hol WG (2004) High resolution crystal structure of a key editosome enzyme from *Trypanosoma brucei*: RNA editing ligase 1. J Mol Biol 343:601–613

Deng J, Ernst NL, Turley S, Stuart KD, Hol WG (2005) Structural basis for UTP specificity of RNA editing TUTases from *Trypanosoma brucei*. EMBO J 24:4007–4017

Drozdz M, Palazzo SS, Salavati R, O'Rear J, Clayton C, Stuart K (2002) TbMP81 is required for RNA editing in *Trypanosoma brucei*. EMBO J 21:1791–1799

Eperon IC, Janssen JW, Hoeijmakers JH, Borst P (1983) The major transcripts of the kinetoplast DNA of *Trypanosoma brucei* are very small ribosomal RNAs. Nucleic Acids Res 11:105–125

Ernst NL, Panicucci B, Igo RP Jr, Panigrahi AK, Salavati R, Stuart K (2003) TbMP57 is a 3′ terminal uridylyl transferase (TUTase) of the *Trypanosoma brucei* editosome. Mol Cell 11:1525–1536

Estevez AM, Simpson L (1999) Uridine insertion/deletion RNA editing in trypanosome mitochondria – a review. Gene 240:247–260

Feagin JE, Stuart K (1988) Developmental aspects of uridine addition within mitochondrial transcripts of *Trypanosoma brucei*. Mol Cell Biol 8:1259–1265

Feagin JE, Jasmer DP, Stuart K (1987) Developmentally regulated addition of nucleotides within apocytochrome b transcripts in *Trypanosoma brucei*. Cell 49:337–345

Feagin JE, Abraham JM, Stuart K (1988) Extensive editing of the cytochrome c oxidase III transcript in *Trypanosoma brucei*. Cell 53:413–422

Gott JM, Emeson RB (2000) Functions and mechanisms of RNA editing. Annu Rev Genet 34:499–531

Gray MW (1994) RNA. Pan-editing in the beginning. Nature 368:288

Gray MW (2003) Diversity and evolution of mitochondrial RNA editing systems. IUBMB Life 55:227–233

Hajduk SL, Harris ME, Pollard VW (1993) RNA editing in kinetoplastid mitochondria. Faseb J 7:54–63

Hajduk SL, Adler B, Madison-Antenucci S, McManus M, Sabatini R (1997) Insertional and deletional RNA editing in trypanosome mitochondria. Nucleic Acids Symp Ser, pp 15–18

Hensgens LA, Brakenhoff J, De Vries BF, Sloof P, Tromp MC, Van Boom JH, Benne R (1984) The sequence of the gene for cytochrome c oxidase subunit I, a frameshift containing gene for cytochrome c oxidase subunit II and seven unassigned reading frames in *Trypanosoma brucei* mitochondrial maxi-circle DNA. Nucleic Acids Res 12:7327–7344

Hong M, Simpson L (2003) Genomic organization of *Trypanosoma brucei* kinetoplast DNA minicircles. Protist 154:265–279

Kable ML, Seiwert SD, Heidmann S, Stuart K (1996) RNA editing: a mechanism for gRNA-specified uridylate insertion into precursor mRNA. Science 273:1189–1195

Kleisen CM, Weislogel PO, Fonck K, Borst P (1976) The structure of kinetoplast DNA. 2. Characterization of a novel component of high complexity present in the kinetoplast DNA network of *Crithidia luciliae*. Eur J Biochem 64:153–160

Klingbeil MM, Englund PT (2004) Closing the gaps in kinetoplast DNA network replication. Proc Natl Acad Sci USA 101:4333–4334

Knight SW, Bass BL (2002) The role of RNA editing by ADARs in RNAi. Mol Cell 10:809–817

Koslowsky DJ, Bhat GJ, Perrollaz AL, Feagin JE, Stuart K (1990) The MURF3 gene of *T. brucei* contains multiple domains of extensive editing and is homologous to a subunit of NADH dehydrogenase. Cell 62:901–911

Liu B, Liu Y, Motyka SA, Agbo EE, Englund PT (2005) Fellowship of the rings: the replication of kinetoplast DNA. Trends Parasitol 21:363–369

Luciano DJ, Mirsky H, Vendetti NJ, Maas S (2004) RNA editing of a miRNA precursor. RNA 10:1174–1177

Madison-Antenucci S, Grams J, Hajduk SL (2002) Editing machines: the complexities of trypanosome RNA editing. Cell 108:435–438

Marchal C, Ismaili N, Pays E (1993) A ribosomal S12-like gene of *Trypanosoma brucei*. Mol Biochem Parasitol 57:331–334

Matthews KR (2005) The developmental cell biology of *Trypanosoma brucei*. J Cell Sci 118:283–290

Michels PA, Bringaud F, Herman M, Hannaert V (2006) Metabolic functions of glycosomes in trypanosomatids. Biochim Biophys Acta 1763:1463–1477

Ochsenreiter T, Hajduk SL (2006) Alternative editing of cytochrome c oxidase III mRNA in trypanosome mitochondria generates protein diversity. EMBO Rep 7:1128–1133

Ochsenreiter T, Cipriano M, Hajduk SL (2007) KISS: the kinetoplastid RNA editing sequence search tool. RNA 13:1–4

Panigrahi AK, Allen TE, Stuart K, Haynes PA, Gygi SP (2003a) Mass spectrometric analysis of the editosome and other multiprotein complexes in *Trypanosoma brucei*. J Am Soc Mass Spectrom 14:728–735

Panigrahi AK, Schnaufer A, Ernst NL, Wang B, Carmean N, Salavati R, Stuart K (2003b) Identification of novel components of *Trypanosoma brucei* editosomes. RNA 9:484–492

Panigrahi AK, Ernst NL, Domingo GJ, Fleck M, Salavati R, Stuart KD (2006) Compositionally and functionally distinct editosomes in *Trypanosoma brucei*. RNA 12:1038–1049

Parsons M (2004) Glycosomes: parasites and the divergence of peroxisomal purpose. Mol Microbiol 53:717–724

Payne M, Rothwell V, Jasmer DP, Feagin JE, Stuart K (1985) Identification of mitochondrial genes in *Trypanosoma brucei* and homology to cytochrome c oxidase II in two different reading frames. Mol Biochem Parasitol 15:159–170

Pollard VW, Hajduk SL (1991) *Trypanosoma equiperdum* minicircles encode three distinct primary transcripts which exhibit guide RNA characteristics. Mol Cell Biol 11:1668–1675

Pollard VW, Rohrer SP, Michelotti EF, Hancock K, Hajduk SL (1990) Organization of minicircle genes for guide RNAs in *Trypanosoma brucei*. Cell 63:783–790

Pollard VW, Harris ME, Hajduk SL (1992) Native mRNA editing complexes from *Trypanosoma brucei* mitochondria. EMBO J 11:4429–4438

Read LK, Wilson KD, Myler PJ, Stuart K (1994) Editing of *Trypanosoma brucei* maxicircle CR5 mRNA generates variable carboxy terminal predicted protein sequences. Nucleic Acids Res 22:1489–1495

Riou G, Delain E (1969) Electron microscopy of the circular kinetoplastic DNA from *Trypanosoma cruzi*: occurrence of catenated forms. Proc Natl Acad Sci USA 62:210–217

Salavati R, Ernst NL, O'Rear J, Gilliam T, Tarun S Jr, Stuart K (2006) KREPA4, an RNA binding protein essential for editosome integrity and survival of *Trypanosoma brucei*. RNA 12:819–831

Schaub M, Keller W (2002) RNA editing by adenosine deaminases generates RNA and protein diversity. Biochimie 84:791–803

Schnaufer A, Panigrahi AK, Panicucci B, Igo RP Jr, Wirtz E, Salavati R, Stuart K (2001) An RNA ligase essential for RNA editing and survival of the bloodstream form of *Trypanosoma brucei*. Science 291:2159–2162

Schnaufer A, Ernst NL, Palazzo SS, O'Rear J, Salavati R, Stuart K (2003) Separate insertion and deletion subcomplexes of the *Trypanosoma brucei* RNA editing complex. Mol Cell 12:307–319

Schumacher MA, Karamooz E, Zikova A, Trantirek L, Lukes J (2006) Crystal structures of *T. brucei* MRP1/MRP2 guide-RNA binding complex reveal RNA matchmaking mechanism. Cell 126:701–711

Shikanai T (2006) RNA editing in plant organelles: machinery, physiological function and evolution. Cell Mol Life Sci 63:698–708

Simpson L, Sbicego S, Aphasizhev R (2003) Uridine insertion/deletion RNA editing in trypanosome mitochondria: a complex business. RNA 9:265–276

Simpson L, Aphasizhev R, Gao G, Kang X (2004) Mitochondrial proteins and complexes in *Leishmania* and *Trypanosoma* involved in U-insertion/deletion RNA editing. RNA 10:159–170

Singh M, Emeson RB (2001) Adenosine to inosine RNA editing: substrates and consequences. In: Bass BL (ed) RNA editing. Frontiers in Molecular Biology, vol 34. Oxford University Press, London, pp 109–138

Sommer B, Kohler M, Sprengel R, Seeburg PH (1991) RNA editing in brain controls a determinant of ion flow in glutamate-gated channels. Cell 67:11–19

Souza AE, Myler PJ, Stuart K (1992) Maxicircle CR1 transcripts of *Trypanosoma brucei* are edited and developmentally regulated and encode a putative iron-sulfur protein homologous to an NADH dehydrogenase subunit. Mol Cell Biol 12:2100–2107

Souza AE, Shu HH, Read LK, Myler PJ, Stuart KD (1993) Extensive editing of CR2 maxicircle transcripts of *Trypanosoma brucei* predicts a protein with homology to a subunit of NADH dehydrogenase. Mol Cell Biol 13:6832–6840

Stagno J, Aphasizheva I, Rosengarth A, Luecke H, Aphasizhev R (2007) UTP-bound and Apo structures of a minimal RNA uridylyltransferase. J Mol Biol 366:882–899

Steinert M, Assel S (1975) Large circular mitochondrial DNA in *Crithidia luciliae*. Exp Cell Res 96:406–409

Stuart K, Panigrahi AK (2002) RNA editing: complexity and complications. Mol Microbiol 45:591–596

Stuart K, Allen TE, Kable ML, Lawson S (1997) Kinetoplastid RNA editing: complexes and catalysts. Curr Opin Chem Biol 1:340–346

Stuart KD, Schnaufer A, Ernst NL, Panigrahi AK (2005) Complex management: RNA editing in trypanosomes. Trends Biochem Sci 30:97–105

Sturm NR, Simpson L (1990) Kinetoplast DNA minicircles encode guide RNAs for editing of cytochrome oxidase subunit III mRNA. Cell 61:879–884

van der Spek H, Arts GJ, Zwaal RR, van den Burg J, Sloof P, Benne R (1991) Conserved genes encode guide RNAs in mitochondria of *Crithidia fasciculata*. EMBO J 10:1217–1224

Wedekind JE, Dance GS, Sowden MP, Smith HC (2003) Messenger RNA editing in mammals: new members of the APOBEC family seeking roles in the family business. Trends Genet 19:207–216

WHO (2006) African trypanosomiasis (sleeping sickness). Fact sheet N°259. WHO, Geneva

Evolutionary Aspects of RNA Editing

Dave Speijer

> *"Complex, statistically improbable things are by their nature more difficult to explain than simple, statistically probable things."*
>
> *Richard Dawkins*

Abstract   RNA editing is the sequence alteration of RNA molecules by nucleotide insertion/deletion or conversion mechanisms. In this chapter, I describe how the different forms of RNA editing may have evolved from pre-existing activities. It appears that repeated and widespread independent evolution of RNA editing occurred. The diversity in origins seems to be mirrored in the range of possible functions of editing:

(1) Multiple proteins could be encoded by one gene. Different editing patterns would generate several proteins from one gene. Conversion editing in vertebrate mRNAs seems to be an instance of such an adaptive function. (2) RNA editing could provide organisms with an extra level of regulation of gene expression, and

Academic Medical Center (AMC), Department of Medical Biochemistry, University of Amsterdam, Meibergdreef 15, 1105 AZ Amsterdam, The Netherlands; d.speijer@amc.uva.nl

H.U. Göringer (ed.), *RNA Editing. Nucleic Acids and Molecular Biology 20*
© Springer-Verlag Berlin Heidelberg 2008

indications for this function are seen in most RNA editing forms. (3) Editing could serve as a defence against viruses and transposons. This could be another role of editing of vertebrate mRNAs. (4) Editing might counteract mutations which have occurred in the genome. These could occur particularly in organellar genomes, when selective pressures are absent. This role may be the raison d'être of mitochondrial tRNA editing. (5) RNA editing could offer the possibility to retain 'difficult' coding sequences, and such a function might be performed by mitochondrial RNA editing in myxomycetes. (6) Last but not least, RNA editing could speed up evolution by creating higher amounts of genetic variation over a shorter period of time. For its function, this model relies heavily on an analogy with splicing, where the possibility of domain shuffling has been invoked as a functional advantage.

All these explanations seem not to suffice for kinetoplastid panediting, the most complex and extensive form of RNA editing known. In this case, I propose that the original advantage was found in the gene fragmentation it entails, protecting against loss of temporarily non-expressed mt genes during periods of intense intraspecific competition. Present-day kinetoplastid editing, however, reflects the effects of a long history of opposing selective forces obscuring its evolutionary origin.

1   Introduction

RNA editing, the sequence alteration of RNA molecules which was first discovered in the mitochondria (mt) of trypanosomes in 1986 (Benne et al. 1986), has been found in many different organisms. The evolutionary histories and possible roles of the different forms of RNA editing seem to be as diverse. Here, I will deal with the evolutionary aspects of most forms of RNA editing, going from relatively simple forms and simple evolutionary explanations to the most complex, kinetoplastid panediting (see chapter by Torsten Ochsenreiter and Steve Hajduk, and chapter by Jason Carnes and Ken Stuart, this volume). Darwinian evolution can be seen as the continuous interplay between chance, generating diversity, and selection, restricting it. The relative contribution of either to the development of a present-day biological process can vary. By looking at the different RNA editing forms in this context, I hope to highlight the great diversity mentioned above.

2   Evolutionary Models

In 1993, Patrick Covello and Michael Gray (Covello and Gray 1993) published a three-step model explaining the evolution of all RNA editing forms. The first step is the acquisition of RNA editing activity by pre-existing enzymes. Next, mutation at 'editable' nucleotide positions in the genome occurs and, finally, editing becomes essential for survival upon fixation by genetic drift. Only in this last step of their proposed mechanism does natural selection come into play: the RNA editing activity is maintained because, to make functional RNAs, it has become indispensable.

Thus, the 'chance' contribution in their model is dominant. This is clear from the fact that the organism with RNA editing capacity does not have an adaptive advantage in comparison to the 'original' organism. Aptly, the model is sometimes referred to as the 'they got stuck with it' hypothesis. All other models invoke evolutionary advantage(s) of the RNA editing capacity as the reason for its retention in present-day organisms. Two types of advantage can be distinguished, though the distinction is artificial and far from absolute. Direct adaptive advantages refer to changes in the organism which immediately bestow improved fitness (e.g. a change in a receptor protein so that a parasite can no longer enter the host). Indirect adaptive advantages, on the other hand, are changes in the organism which do not necessarily lead to improved fitness as such, though preferably they should not compromise individual fitness, but which lead to higher 'evolvability'. 'Evolvability' refers to changes in the organism that, though possibly without any impact whatsoever on the organism's direct survival, allow descendants of the organism to spread more successfully. A few examples will highlight the salient points of these kinds of genetic changes.

To explain the widespread occurrence of splicing in eukaryotes, it has been proposed that (one of) its evolutionary benefit(s) lies in the fact that it allows functional protein domains to be exchanged, making very rapid modular protein evolution possible. The development of sexual exchange of genetic material in all its different forms opened up far more efficient methods of recombination. This gives rise to much more complex samples of genetic combinations for natural selection to choose from (Hoekstra 2005). Thus, one could divide the proposed adaptive functions of RNA editing into two groups, based on whether a direct survival benefit or a higher level of 'evolvability' predominates. It should be stressed again that the distinction is artificial: examples of RNA editing evolution 'having a bit of both' will highlight this further. The following list of proposed advantages for RNA editing starts with direct benefits and leads on to 'evolvability' advantages.

1. Multiple proteins could be encoded by one gene. Different editing patterns (including absence of editing) would allow the generation of several proteins from only one gene. Especially organisms in which genome space is limited, such as viruses (although the recent discovery of the 1.2-Mb Mimivirus gives pause for thought; Raoult et al. 2004), would be expected to make use of the possibilities offered by RNA editing in this respect.
2. RNA editing could provide organisms with an extra level of regulation of gene expression. As editing is an essential step in the maturation of the RNAs which are subject to this process, it is hard to imagine the organisms not developing mechanisms to regulate gene expression at this point.
3. RNA editing might function as an efficient defence against viruses and transposons (Grivell 1993). Genetic parasites would be made harmless when their RNAs would be rendered meaningless by editing.
4. Another advantage of the RNA editing potential could be that it offers the possibility to retain 'difficult' coding sequences. It is imaginable that a relatively unsophisticated replication or transcription machinery would encounter problems

with certain sequences. These sequences could then be generated by RNA editing (Grivell 1993).

5. Editing is sometimes seen as counteracting mutations which have occurred in the genome. These could occur especially in organellar genomes, accumulating during periods when organellar function is not, or only partially, needed, and selective pressures thus absent (see, e.g. Cavalier-Smith 1997).

6. It has been proposed that RNA editing can speed up evolution by creating higher amounts of genetic variation over a shorter period of time. Indeed, edited sequences seem to have a higher rate of molecular evolution (Landweber and Gilbert 1993). This model relies heavily on an analogy with splicing, where the possibility of domain shuffling (cf. above) has been invoked as functional advantage. This would strongly favour the retention of the splicing machinery.

7. For kinetoplastid panediting, it has been proposed that the advantage is provided by the gene fragmentation it entails, protecting against loss of temporarily non-expressed mt genes during periods of intense intraspecies competition (Speijer 2006 and see below). This would force organisms to retain genes which are not under selective pressure during prolonged periods of growth and division, thus retaining a higher 'ecological' flexibility which would, e.g. enable them to evolve complex (parasitic) lifecycles.

Regarding the concept of evolvability, it should be stressed that the apparent foresight of the examples mentioned is indeed only 'apparent'. If a mechanism giving rise to an increase in genetic flexibility would be really detrimental to the organism, severely hampering its chances to generate offspring, then all putative benefits for future generations would not lead to its retention in the species. Such a mechanism should at least be (almost) neutral for the organism to give rise to a line in which 'evolution speeds up'. Let us look at most of the known editing mechanisms in the light of all these different proposals to explain their occurrence, starting with the most 'simple'.

3 RNA Editing in Paramyxoviruses

RNA editing in paramyxoviruses is the co-transcriptional process leading to extra inserted G residues in the viral P mRNA, which are derived from the action of a stuttering polymerase (Vidal et al. 1990). This type of editing falls into the category of co-transcriptional RNA alteration processes which historically have been considered forms of editing (Hausmann et al. 1999). Another example of co-transcriptional viral editing is found in the insertion of an A residue in the G(lyco)P(rotein) mRNA of Ebola viruses (Volchkov et al. 1995), switching expression from the encoded (non-structural) secreted glycoprotein to that of the structural virion glycoprotein. This editing also seems to be due to a stuttering polymerase. These editing capabilities are retained by the organisms in order to make multiple proteins with clearly different functions in the lifecycle of these viruses. As a further bonus, the expression of the various P or GP gene products, the relative proportions of which seem to be critical, can thus be tightly regulated.

4 tRNA Editing in Mitochondria

The mt tRNA editing examples (see also chapter by Juan Alfonzo, this volume) seem to be quite compatible with the Covello and Gray model. Evolutionary fixation of editing has been nicely described for marsupial mt tRNA editing in Börner and Pääbo (1996). First, a T to C transition in the gene for tRNA[Asp] occurred. It changed the anticodon of this tRNA, allowing recognition of GGY codons and a concomitant shift in charging from aspartic acid to glycine. This turned out to be not lethal to the organism, presumably because a pre-existing deaminase activity converted this nt in the tRNA back into U, and because the presence of multiple copies of the non-mutated tRNA in the mt genome allowed this activity to become more efficient. This resulted in an editing activity changing about 50% of the mutated tRNA back into the original tRNA[Asp]. A situation in which two of the GGN glycine codons were recognized by two different tRNAs was the result: GGY codons were decoded by the unedited tRNA[Asp] and the original tRNA[Gly]. Restoration of the original situation became impossible because of a second mutation making the RNA editing process indispensable. This mutation occurred in the tRNA[Gly] gene, limiting the tRNA[Gly]'s decoding capacity to the two GGR codons, in the process making both forms (edited and unedited) of tRNA[Asp] essential for survival. In plant organellar editing, a similar situation exists: in this case, the abundance of editing sites makes it impossible to revert to a situation where loss of editing activity would not be lethal. The mt tRNA editing activity in marsupials has not been shown to have any function apart from the repair of a genomic mutation at the level of the RNA. An overview of mt tRNA editing is given in Table 1.

All instances of tRNA editing seem to have no function apart from 'repair' of the encoded tRNAs which have accumulated mutations impairing proper functioning of the tRNA in translation and, thus, seem to conform to the original Covello and Gray model. tRNA editing independently evolved many times in the mitochondria of diverse eukaryotes, co-opting mostly different pre-existing enzymatic activities seemingly without any direct benefit to the organism. The widespread occurrence and independent evolution of organellar RNA editing is illustrated in Fig. 1.

5 RNA Editing in Plant Organelles

Plant pyrimidine (C to U and U to C) editing is very widespread in plant organelles, especially in mitochondria (see chapter by Mizuki Takenaka and colleagues, and chapter by Masahiro Sugiura, this volume, and below). As the raison d'être for the existence of RNA editing in plant organelles, the 'one gene, multiple proteins' explanation has often been invoked. Is this correct? There are several observations which make it unlikely.

First of all, the large majority of plant editing events changes transcripts 'back' into the form which is conserved over distantly related species, giving the impression these

Table 1 tRNA editing in mitochondria

Organism	Site	Pre-existing enzyme	Remarks	Ref.
Marsupials	Anticodon	Deaminase	See text	Börner and Pääbo (1996)
Leishmania tarentolae	Anticodon	Deaminase	*Imported* tRNA[Trp]	Kapushoc et al. (2000)
Acanthamoeba castellanii	5′ acceptor stem	?	12 tRNAs	Lonergan and Gray (1993)
Chytridiomycete fungi	5′ acceptor stem	?	Evolved at least twice	Laforest et al. (1997, 2004)
Naegleria Gruberi	5′ acceptor stem	?	Lobose amoeba (see Fig. 1); predicted only	
Land snails	3′ acceptor stem	Poly(A) polymerase[a]		Yokobori and Pääbo (1995a)
Squids	3′ acceptor stem	Poly(A) polymerase[a]		Tomita et al. (1996)
Gallus gallus	3′ acceptor stem	Poly(A) polymerase[a]		Yokobori and Pääbo (1997)
Ornithorhyncus anatinus	3′ acceptor stem	CCase[b]		Yokobori and Pääbo (1995b)
Lithobius forficatus	3′ acceptor stem	RdRp?[c]	22 cases predicted, eight confirmed	Lavrov et al. (2000)
Seculamonas ecuadoriensis	3′ acceptor stem	RdRp?[c]	Jakobid (see Fig. 1)	Leigh and Lang (2004)

[a] Examples of the very efficient use of animal mt genome coding capacity (e.g. in *G. gallus*, mt tRNA[Tyr] and tRNA[Cys] overlap by a single G residue; see also below), closely resembling 3′ polyadenylation of some vertebrate mt mRNAs where UAA stop codons result from A addition(s) at the 3′ end (Ojala et al. 1981)
[b] (modified?) tRNA nucleotidyl transferase, involved in 3′ CCA addition to tRNAs
[c] RNA-*dependent* RNA polymerase. N.B. poly(A)polymerase and CCase are template-independent

organelles use editing only as a repair mechanism and that plants would be better off starting out with the 'correct' DNA sequence (Tsudzuki et al. 2001). Pyrimidine interconversions observed in plant mitochondria do give rise to different transcripts associated with polysomes in the process of being translated (Lu and Hanson 1996). However, studies with antibodies against the unedited ATP6 protein showed translation products from unedited mRNAs to be very unstable (Lu and Hanson 1994). Even when products of unedited transcripts can be detected, they are absent from complexes they should be part of, despite the fact that, compared to the proteins as encoded by edited transcripts, only minor changes are present. This is illustrated by the product of unedited RPS12 RNA, which is not detectable in plant mitoribosomes (Hanson 1996). The role of proteins encoded by (partially) unedited messengers, if any, thus remains unclear. Nevertheless, it seems likely that plants do use at least some of the editing instances to control organellar gene expression. An example is found in the tobacco plastid ndhD transcript in which the initiation codon is generated by a C to U conversion (Hirose and Sugiura 2001). The transcript is (partially) edited in leaves

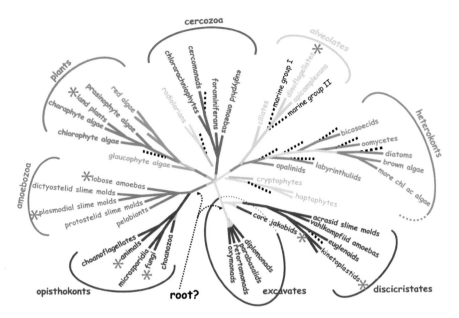

Fig. 1 Repeated and widespread independent evolution of organellar RNA editing in eukaryotes. The unrooted tree has been adapted from Baldauf (2003). *Red asterisks* indicate the acquisition of RNA editing in the clade. Comparative studies demonstrate multiple independent instances of organellar RNA editing acquisition also within clades: so far, it has been observed in fungi, animals, plasmodial slime moulds, and kinetoplastids (see text)

containing active NADH dehydrogenase and not edited in root, which does not make the active complex. This regulation could be achieved by, for example, controlling the availability of one of the hundreds of pentatricopeptide repeat (PPR) motif-containing proteins implicated in the specification of editing sites (Shikania 2006). It should be noted that the *Arabidopsis* transcript has four additional editing sites which all have to be converted before the transcript encodes the correct protein, and that wheat, rice and maize all encode the start AUG (rather than ACG in tobacco) directly in the plastid genome (Tsudzuki et al. 2001). Again, these observations seem to be in agreement with an editing mechanism that simply 'took over' by chance without any direct benefit to the organism, editing only subsequently being used as an extra level of gene expression control and, even then, only in a minority of cases.

One can also compare the overall evolution of genomic architecture in organelles of different eukaryotes to see what can be learned about the evolution of editing in different organelles. This comparison was performed by Lynch et al. (2006). They show quite convincingly that the genomic development of mitochondria (and chloroplasts) can be understood as the evolutionary result of an interplay of two *non-adaptive forces only*: random genetic drift and mutation pressure. Comparing the mt mutation rate (measured as silent site divergence rate) between different phylogenetic groups, they found that mammalian mitochondria have the highest and plant mitochondria by far the lowest mutation rate (i.e. a hundredfold lower). The mutation rates in the mitochondria of unicellular organisms and plant chloroplasts fall between these two extremes. It is clear from their analysis that genome size, amount of intergenic DNA, amount of introns and, last but not least, amount of editing in organelles are *inversely* related to the local mutation rate. The highly reliable plant mt replication allows these organelles to retain large mt genomes with rampant editing without direct evolutionary benefits or even a (slight) selective disadvantage: 'restorative' mutations, rendering editing superfluous, occur four times as fast as the neutral rate would predict (Shields and Wolfe 1997). Mammalian mitochondria, on the other hand, with their much higher mutation rate, have to counteract this error rate by reducing their mt DNA to the absolute minimum, with very limited editing only if allowing further reduction of overall genome size (see above). Superficially, this might seem to contradict the 'non-adaptive' Covello and Gray model, because one might conclude that as a genomic repair mechanism, editing should be most abundant in organelles with the highest mutation rate. This is, however, not correct. The generation of mutations is not 'limiting' (the overwhelming majority of non-silent mutations simply die out) but the possibility of retaining mutations in the absence of an adaptive advantage is. The mutation rate thus limits the amount of non-beneficial 'weight' a genome can carry (compare, e.g. the viral error catastrophe concept). The observed editing patterns thus seem to be in agreement with a non-adaptive model, with the amount of pyrimidine interconversion and tRNA editing observed in chloroplasts and animal, unicellular and plant mitochondria being determined by the relative fidelity of the operating replication system.

Although Lynch et al. (2006) do not mention this possibility, it is tempting to extend their argument: in mt systems where the amount of editing is much higher than expected on the basis of the local mutation rate, real adaptive advantages must (have) be(en) present. Assuming a 'normal' mutation rate for the mitochondria of unicellular organisms, only a few cases of mt editing seem to be so massive that this could be indicative of real adaptive advantages: possibly, mixed substitution editing in dinoflagellates (Lin et al. 2002), mixed editing in myxomycetes (see chapter by Jonatha Gott and Amy Rhee, this volume) and, especially, U-insertion/deletion editing in kinetoplastids. Before discussing the latter two in more detail, examples of less extensive editing should be looked at. These forms of non-organellar conversion editing have clear adaptive advantages: anti-viral properties and multiple protein production.

6   RNA Editing in Cytoplasm and Nucleus

The instances of conversion editing in vertebrate mRNAs do demonstrate contributions beyond 'genome repair' of RNA editing. Cases of RNA editing which have been reported include C to U editing in the mammalian nuclear apolipoprotein B (apoB) mRNA (Davidson 1993), in the Neurofibromatosis type-1 (NF1) mRNA (Skuse et al. 1996) as well as in several viral RNA and DNA molecules (e.g. in HIV; see Turelli and Trono 2005), and A to I editing in mRNAs encoding vertebrate and insect neurotransmitter receptor subunits (Sommer et al. 1991; Palladino et al. 2000) as well as in viral antigenomic RNAs of human hepatitis delta virus (Polson et al. 1996).

In apoB mRNA, a glutamine codon (CAA) is changed into a stop codon (UAA). The edited mRNA encodes the 48-kDa N-terminal part of the 100-kDa apoB protein encoded by the unedited mRNA. The editing activity is tissue-specific, normally occurring only in the intestine (and in the liver of rodents), and physiologically highly relevant because while both proteins bind lipids to form lipoprotein particles, the ones containing the 48-kDa apoB protein (chylomycrons) are involved in the transport of dietary fat, whereas 100-kDa apoB protein is the major protein component of the low-density lipoprotein particles, the 'bad' cholesterol (Davidson 1993). NF1 mRNA editing is found in certain human tumours. The C to U conversion changes an arginine codon (CGA) into a stop codon (UGA). The truncated product presumably loses its GTPase activity and so does not convert the proto-oncogene product RAS into the inactive (GDP-bound) form. Thus, editing would be selected for during, and contribute to, tumorigenesis. These two instances of C to U conversion are catalyzed by the cytidine deaminase APOBEC-1, the founding member of a large family of C-deaminases with different specificities, which also contains activation-induced deaminase, AID, and 10 other (predicted) APOBEC homologues in humans (Turelli and Trono 2005). AID is involved in class switch recombination and somatic hypermutation (though presumably not as an RNA deaminase), processes leading to the production of high-affinity antibodies by B-cells. At least some of the other APOBEC enzymes are also involved in immune reactions but, in their case, mostly by direct C-deaminase activity on (intermediate) DNA and RNA forms of human viruses, such as HIV (Bishop et al. 2004). However, not all anti-viral activity of the protein family is due to the deaminase activity. As a countermeasure, the HIV-encoded Vif protein specifically targets APOBEC3G and APOBEC3F for proteosomal destruction. Comparing evolution of the family between rodents and humans, diversification seems to have been rapid, especially in humans. This could be related to differences in the ongoing arms' race between viruses and retrotransposons, and their human or rodent hosts (Turelli and Trono 2005). All these homologous zinc-containing proteins most likely have evolved from an ancestor involved in pyrimidine metabolism (Gerber and Keller 2001) which also gave rise to the family of A to I deminases mentioned above (the ADAR enzymes; see also chapter by Michael Jantsch and Marie Öhman, this volume).

In the case of A-deaminase activity, the evolution and functional diversification are strikingly similar to what we observed in the C-deaminase family. A to I editing is found in mRNAs encoding vertebrate neurotransmitter receptor subunits (Sommer et al. 1991). It has been reported for rodents and humans in mRNAs encoding subunits of the glutamate receptor, mediating fast excitatory responses in the central nervous system. These receptors function as glutamate-gated cation channels which open in response to L-glutamate binding. There are three types, called AMPA, NMDA and KA responsive channels (based on agonist sensitivity; see Seeburg 1996). The different sensitivities result from the hetero- or homomeric composition of these multimeric channels, which are built from combinations of gene products encoded by the GluR gene family. The variety in response is further increased by A to I editing, as initially discovered by Seeburg and colleagues (Sommer et al. 1991): e.g. in GluR-B (and in

GluR-5 and 6) mRNA, a CAG codon in exon 11 is changed into a CIG codon which is read as CGG by the translational machinery. This changes the encoded glutamine (Q) into arginine (R) at these so-called Q/R sites. Editing efficiencies differ for the different transcripts (from almost 100% change in GluR-B to 40% in GluR-5 mRNAs; Smith and Snowden 1996). In GluR-B, C and D mRNAs, which all encode subunits of AMPA-sensitive channels, an additional A to I editing site is found in exon 13 changing AGA into IGA (GGA). This arginine to glycine codon change is known as the R/G site. In the GluR-6 transcript, the so-called I/V and Y/C sites also result from editing. The channels are profoundly influenced by these changes: e.g. the Q/R editing places an arginine in a transmembrane domain affecting calcium permeability. A to I editing was also found in the serotonin 2c receptor mRNA. Other human targets of ADARs have recently been identified: although none of these transcripts encode receptors, half of them were strongly expressed in the CNS (Levanon et al. 2005). All the endogenous A to I targets so far identified in rodents, squid, teleost fish (Kung et al. 2001) and *Drosophila* encode proteins functioning in the nervous system. As editing diversity combines with gene (family) diversity, an enormous potential of different specificities and responses could be envisaged, which is of course of great value during evolution, development and functioning of the animal brain. Similarly to the APOBEC family, however, endogenous transcripts are not the only targets. It has been suggested that ADARs also evolved for their protective function against viruses (Scott 1995). That they recognize viral targets is illustrated by the fact that A to I conversion occurs in the human hepatitis delta virus in 20–50% of the viral antigenomic RNAs at the so-called amber/W site, although in this case the virus is not the hapless victim of deaminase activity. Editing changes a stop codon (UAG) into a tryptophane codon (UIG), resulting in a 19 amino acid extension of the reading frame. The shorter protein (hepatitis delta antigen) encoded by the mRNAs transcribed from the unedited template is involved in viral replication, while the longer version is needed for packaging and inhibits replication (Bass 2002). The longer protein actively inhibits editing activity in a negative feedback loop, so that replication will not be inhibited too strongly. From the fact that the respective deaminase activities are involved in the arms' race with pathogens and are used in, e.g. GluR and ApoB editing, giving rise to protein products with physiologically relevant differences, it seems logical to conclude that natural selection actively favoured those organisms which gained (diversification of) deaminase editing capacity.

7 RNA Editing in Myxomycetes

Returning to extensive organellar editing, adaptive advantages, if any, are less clear in this case. In the myxomycete *Physarum polycephalum*, four different forms of editing are operational in the mitochondrion, even working on one transcript: cytochrome c oxidase subunit I (cox1) mRNA undergoes 59 C insertions, one U insertion, three different dinucleotide (CU, GU and UA) insertions, and four C to U base conversions. As seen in this example, editing of mt RNAs in this organism predominantly concerns insertion of single C residues, while less frequently, single Us or dinucleotides (only 19 reported instances, so far) are inserted (Gott et al. 1993; Miller et al. 1993; Visomirski-Robic and Gott 1995; Byrne and Gott 2004). The insertion of single residues at 'regularly' spaced intervals in this and other transcripts results in the removal of multiple gene-encoded frameshifts, making this form of editing a crucial step in gene expression, like its kinetoplastid counterpart (see below). Different from kinetoplastids, however, insertion of residues has also been observed in mt tRNAs and rRNAs (Miller et al. 1993). No consensus sequences defining the hundreds of insertion sites seem to be present but they usually follow a purine/pyrimidine dinucleotide, and are found mostly in third codon positions. Myxomycete insertional editing is distinct from uridine insertion/deletion editing in kinetoplastids: the edited sites have dissimilar patterns, the identity of nucleotides involved is different, myxomycetes apparently lack gRNA-like template molecules, and most importantly, all insertional editing seems to be co-transcriptional. This was shown by analysis of mRNAs obtained under conditions of stalled RNA polymerization, due to limiting concentrations of nucleotides added to isolated mitochondria (Visomirski-Robic and Gott 1997; Byrne and Gott 2004).

A phylogenetic analysis of cox1 genes and transcripts in different myxomycetes allows a reconstruction of the most likely evolutionary history of insertional editing in myxomycetes (Horton and Landweber 2000). Presumably, this form of editing started with an ancestral U-addition capacity only, followed by the acquisition of (other factors allowing) C addition and dinucleotide addition respectively. This is not unlike the situation in kinetoplastids, where apart from the TUTase for U addition or the U-exonuclease for U deletion, the religation of the 5' and 3' parts of the mRNA is also catalyzed by different, insertion- and deletion-specific ligases. A basic operating mechanism is thus extended to enhance its versatility. The C to U conversion editing, which edits mRNAs post-transcriptionally and thus can be separated from insertional editing activity in *P. polycephalum* extracts (Visomirski-Robic and Gott 1997), clearly evolved independently. The complexity of editing in *P. polycephalum* is illustrated not only by the example of the cox1 transcript given above but also by the fact that even inserted C residues could subsequently be converted into U residues (Byrne and Gott 2004). It goes without saying that all the different editing mechanisms are indispensable for the generation of the 'correct' mt mRNAs. The sheer abundance of editing sites (on average, one in every 25 bases is edited), however, seems to demand additional explanation. Without more information about the evolution and specific mechanisms of RNA editing in this lineage, such an explanation is not easy. It is conceivable that myxomycetal insertion editing is the only example of editing evolved specifically to counteract the effects of a relatively 'deletion-prone' mt polymerase in these organisms. Possibly, the replication machinery encounters problems with certain sequences leading to frequent deletions in the mtDNA. These sequences could then be generated co-transcriptionally (compare model 4 above). In the discussion of pyrimidine interconversion editing in plant organelles, the work by Lynch et al. (2006) was quoted: overall genomic architecture is linked to respective mutation rate. Could this insertional editing be an instance of coevolution with the idiosyncrasies of a specific mt polymerase? The insertional editing level possibly is so high because it has to counteract a relatively high rate of *very specific* mutations only: single nucleotide (and very rarely, double) deletions. This model can explain the following aspects.

1. In most cases, editing follows a purine/pyrimidine dinucleotide sequence: this should be part of the sequence forcing the mt polymerase to make occasional deletions.
2. It fits with the strong preference for *insertion* at the third position of the encoding triplet in mRNAs: only in that position is the identity of the inserted nucleotide in most cases irrelevant to the protein encoded. Thus, third position deletions have a much better change of survival if a 'blind' insertion mechanism is present at the transcriptional level. This also explains the surprising difference with pyrimidine interconversions in plant mitochondria, which occur *least* in the third position. In this case, editing of a first or second position is actively maintained by selective pressure and becomes unnecessary only in the case of a 'restorative' mutation, while third position *conversion* can 'come and go as it pleases', being neutral.
3. The replacement of the ancestral U-insertion mechanism by the very extensive C-insertion mechanism also makes some sense: the identity of the inserted nucleotide is not important, allowing C insertion to take over, possibly as a result of genetic drift only.
4. This type of insertion editing is really different from that observed in kineto-plastids, where transcripts are either not, only very locally, or completely changed by U insertion. If the editing sites in myxomycetes are due only to dele-tions resulting from random occurrence of certain sequences in the DNA, then one would expect editing sites to be 'sprinkled' rather regularly all over their mt genomes. This is indeed observed. The appearance of 'evenly' spaced editing sites again results from the strong preference for their occurrence in third codon positions, with the minimum distance of nine nucleotides possibly due to processivity demands of the RNA polymerase.

However, some questions remain. If the model of Lynch et al. is correct, then despite being relatively 'deletion-prone', the specific mutation rate of mt DNA polymerase from *P. polycephalum* must be relatively low to sustain a genome with such an abundance of RNA editing. Alternatively, myxomycetal insertion editing confers a real adaptive advantage, which has not been identified yet. In the light of this hypothesis, the characterization of mt DNA polymerase from *P. polycephalum* should have high priority.

8   RNA Editing in Kinetoplastids

8.1   *Introduction*

RNA editing was discovered in trypanosomatids, parasitic unicellular organisms belonging to the kinetoplastid order, in 1986 (Benne et al. 1986), and in this instance defined as the process of post-transcriptional sequence alteration via the

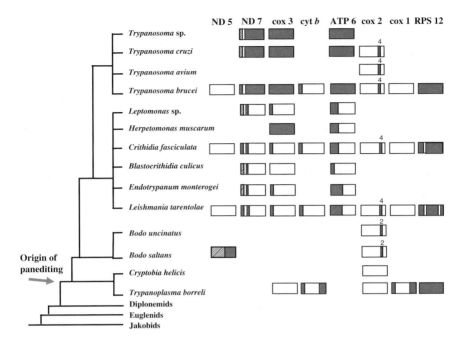

Fig. 2 Evolution of uridine insertion and deletion editing in kinetoplastids. An updated consensus phylogenetic tree of kinetoplastids is given (based on Maslov and Simpson 1994; Arts and Benne 1996; Blom et al. 1998; Simpson et al. 2002). On the *right*, editing patterns of eight mRNAs are shown (not drawn to scale) as far as they are known for each species. The *boxes* indicate (the regions of) the transcripts which are edited (*red*) or unedited (*white*). *Hatched boxes* indicate that the editing status of the segment is not known. Cox II editing has been observed with insertion of either two or four residues, as indicated above the boxes. The inferred stage at which panediting originated is indicated by a *red arrow*

insertion and deletion of uridylate residues at specific sites of mt RNAs. mt RNAs are encoded by the kinetoplast DNA (kDNA) which, in most cases, contains several thousand small minicircles and a few dozen maxicircles (Benne 1993; Simpson et al. 2000; Madison-Antenucci et al. 2002). The maxicircles encode subunits of the respiratory chain complexes and the ribosomal RNAs, while the minicircles encode the so-called guide (g) RNAs containing information for the editing of mt mRNAs encoded by the maxicircles (Blum et al. 1990). The large majority of the maxicircle genes encode RNAs which, without editing, are not translationally competent because they encode frameshifted proteins and/or lack translation initiation codons. A number of transcripts are edited only at the 5′ termini, which results in extra codons for N-terminal amino acids, including an in-frame AUG start codon. Examples of 5′ editing are found in, e.g. cyt*b* mRNA (see, e.g. Shaw et al. 1988; van der Spek et al. 1990) in *Trypanosoma brucei*, *Crithidia fasciculata* and *L. tarentolae* (see Fig. 2). Editing can be so extensive that certain genes are not recognizable as such in the maxicircle of certain trypanosomatids. A good example is the cox3 gene, which can easily be identified in *C. fasciculata* and *L. tarentolae* (the transcripts of which are subject only to 5′ editing). The *T. brucei* cox3 'gene' appears to be absent and can be recognized only after editing over the complete length of the transcript (Feagin et al. 1988). While the G, A and C nucleotides (the GAC sequence) are encoded by the maxicircles, the U sequence is generated by editing: 547 Us are inserted and 41 Us are deleted at 223 sites. More than half of all nucleotides in this transcript are derived from editing. Because these genes can not be identified on the basis of their GAC sequence alone, they are referred to as cryptogenes while this form of extensive editing has been christened panediting. Panediting is common in *T. brucei*, and nine transcripts are edited in this fashion. From the analyses of cDNAs of several partially edited transcripts and RT-PCR experiments with combinations of oligos recognizing edited and unedited sequences, an overall 3′ to 5′ polarity of the editing process could be deduced (Feagin et al. 1988; Koslowsky et al. 1990; van der Spek et al. 1991). In this volume, the precise molecular mechanisms of the editing process are described in the chapter by Jason Carnes and Ken Stuart (also see Cruz-Reyes et al. 2002; Madison-Antenucci et al. 2002 and references therein; Schnaufer et al. 2003).

8.2   Evolution and Age of Kinetoplastid RNA Editing

Information regarding the evolutionary history of the RNA editing process can be obtained from studies with the bodonid species *Trypanoplasma borreli*, which demonstrated the existence of the U-insertion/deletion type of RNA editing also in this organism (Lukeš et al. 1994). This indicates that the process is at least 500–700 million years old, given the estimated time of divergence of the two kinetoplastid suborders (Fernandes et al. 1993). The flagellates (kinetoplastids, diplonemids and euglenoids) seem to form one of the earliest diverging eukaryotic lineages which have mitochondria. Research in *Euglena gracilis* and *Diplonema papillatum* has not yet revealed evidence of the existence of mt RNA editing in the non-kinetoplastid flagellates (Marande et al. 2005). Although this research has been far from extensive, most likely U-insertion/deletion editing evolved only in the kinetoplastid lineage (see Fig. 2).

Panediting occurs not only in *T. brucei* but also in cultivated strains of *L. tarentolae* and *C. fasciculata*, although clearly less frequently. In these organisms, the cox3 and ND7 genes are much less cryptic, the transcripts requiring only limited editing. Interestingly, the potential to edit transcripts encoding proteins which are not necessary during prolonged culture has been lost. When freshly isolated from the host, *L. tarentolae* (LEM 125 strain) does show extensive editing of six transcripts (Maslov et al. 1992; Thiemann et al. 1994). Extensive editing has further been observed in the trypanosomatids *Trypanosoma cruzi* and *Trypanosoma* species E1-CP, infecting fish, and *Herpetomonas muscarum*. The bodonid *T. borreli* also displays extensive editing in the RPS12 RNA; by contrast, cox3 RNA, edited in all trypanosomatids studied so far, is completely unedited in this organism (Maslov and Simpson 1994). Combining the kinetoplastid evolutionary tree with the differences in editing patterns observed (Fig. 2; Landweber and Gilbert 1994;

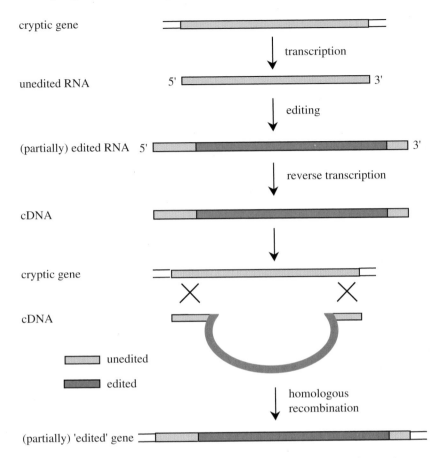

Fig. 3 Model explaining the loss of RNA editing via recombination. A cryptic gene is converted into a (partially) 'edited' gene via homologous recombination with a (partially) 'edited' cDNA. The transcript of the new gene no longer requires extensive editing; 'unedited' sequences are in *green*, 'edited' sequences in *red* (see text for details; taken from Arts and Benne 1996)

Arts and Benne 1996; Blom et al. 1998; Simpson et al. 2002), one can draw conclusions regarding the evolution of kinetoplastid RNA editing. Surprisingly, the distribution of RNA editing patterns in different kinetoplastids clearly indicates panediting to be an ancient trait. Presumably, panediting very quickly followed the evolution of editing itself in the kinetoplastid lineage. One can conclude also that editing becomes progressively less extensive, at times even completely disappearing: e.g. loss of cox3 editing in *Blastocrithidia culicis*. This resembles the current situation in plant organelles where local mutations which do away with the need for editing occur four times faster than the neutral rate (Shields and Wolfe 1997). A possible mechanism for the loss of editing in kinetoplastids was given by Landweber (1992). The model (illustrated in Fig. 3) is based on reverse transcription of (almost completely) edited RNA followed by homologous recombination of the cDNA with the mt DNA. The 5′ and 3′ homology requirements of the cDNA would lead to the occasional retention of the need for 5′ editing of the 'new transcript encoded by the mt DNA following recombination. This is indeed what one observes in, e.g. cox3 in *Endotrypanum monterogei*, *L. tarentolae*, *C. fasciculata* and *Leptomonas* species (compare Figs. 2 and 3). Not surprisingly in the light of its immense complexity and energetic costs, panediting at the current level seems disadvantageous and to be disappearing. One of the driving forces for this could be the loss of gRNA genes. As mentioned above, cultivated strains of *L. tarentolae* clearly show less editing potential upon prolonged culture. In the absence of selective pressures on (some aspects of) mt activity, loss of guide RNAs and loss of heterogeneity of the minicircles seem to occur quite rapidly due to large-scale deletions in, and asymmetrical divisions of, kDNA.

8.3   How Did Kinetoplastid RNA (Pan) Editing Evolve?

Can the Covello and Gray (1993) model account completely for the evolution of kinetoplastid RNA editing? Several considerations make this highly unlikely. As described above, the last step in their model consists of the fixation of RNA editing by genetic drift. The term 'genetic drift' is used here to describe a chance process by which an altered form replaces the original *without* a selective advantage. In the case of kinetoplastid editing, this is rather difficult to conceive (although genetic drift in relatively small populations should not be underestimated): not only all the copies of the mt genome at the time of origin (present-day organisms such as *T. brucei* have 50 maxicircles) but also the population of organisms as a whole have to be replaced without any selective advantage whatsoever, and this hundreds of times! This point will be dealt with below. Natural selection plays a role only in the final step of the Covello and Gray model: the RNA editing activity is maintained because it is required to make functional RNAs. Although this model could indeed apply to rare isolated instances of editing, it is hard to see how it explains the observed rapid acquisition of multiple editing sites in panediting requiring myriads of gRNAs. As stated above in the context of the evolution of mt genomic architecture, the sheer bulk of editing tells us that real adaptive advantages must (have) be(en) present and that the 'they got stuck with it' hypothesis clearly misses a description of the selective pressure(s) responsible for an *active* increase of the editing potential. The distribution of editing patterns excludes the development of panediting as a way to counteract specific mistakes of the kinetoplastid mt DNA polymerase. With regard to the models describing selective advantages, the following can be stated. Multiple proteins could indeed be encoded by one (crypto)gene but, firstly, translatable open reading frames seem to result only from complete editing (see below) and, secondly, no experimental evidence for this model has as yet been reported. Possibly, un- or partially edited transcripts are occasionally used to generate proteins, but this would clearly be a 'late' use of a mechanism rapidly evolving for other reasons. It is also true that editing provides the kinetoplastids with some extra level of regulation of gene expression (e.g. compare editing patterns during different lifecycle stages). However, this can hardly be used to explain the emergence of this highly elaborate system. Both models apply only to simple editing events and can not explain extensive editing. Alternatively, RNA editing could counteract mutations accumulating during periods when mt function is not needed (such as during anaerobiosis with glycosomal activity for energy generation in free-living bodonid ancestors; Cavalier-Smith 1997). The problem with this latter model is clearly that kDNA with multiple gRNAs would become even more mutation-prone because more coding capacity is needed for a given message. To make matters worse, during these periods of absence of selective pressure on mt activity, gRNA loss could be enormous. An alternative explanation was needed (also see Simpson et al. 2002) and has been proposed by Speijer (2006). This model will now be discussed.

8.4 A Possible Explanation for the Spread of Kinetoplastid Panediting

The vast majority of present-day kinetoplastids have a parasitic lifestyle. This means that these organisms grow and divide under varying conditions during their lifecycle (alternating between free-living and host or between different hosts). To understand the emergence of (pan)editing, an example of mt functioning during the lifecycle of trypanosomatids – in this case, *T. brucei* – is described in some detail.

T. brucei infects the mammalian bloodstream and can be taken up by Tsetse flies. It completes its developmental cycle in the mouthparts of the insect and is transmitted to the mammalian host during blood intake by the insect. In the Tsetse fly, a respiratory chain starting with complex II is operational. This implies complete editing of the transcripts for the mt-encoded subunits of complex III and IV which have to be edited (cyt*b* and cox2 and 3 respectively). The transcript of the ATP6 gene product encoding a subunit of the F_oF_1-ATPase should, of course, also be fully edited in the fly. In the long, slender mammalian bloodstream form, the respiratory chain is absent because all energy is generated from glycolysis alone. The F_oF_1-ATPase is still present in the mt inner membrane in this form but, rather than generating ATP from the proton gradient, it generates a proton gradient from ATP, presumably to make protein import into the mt matrix possible under these conditions (Nolan and Voorheis 1992). When the rapidly dividing, long slender form changes into the non-dividing short stumpy form, complex I is formed in the mitochondrion. The mt messengers for ND3, 7, 8 and 9 have to be panedited during this stage. Under these circumstances, ATP is presumably generated by the F_oF_1-ATPase, using the proton gradient generated by complex I (Bienen et al. 1991). When taken up by the fly, only the stumpy form can start the whole cycle again, possibly due to the fact that its metabolic circuitry more closely resembles the one needed in the fly. Although repressed, mt function is clearly not absent in bloodstream parasites. A very extensive description of the energy metabolism during the different lifecycle stages can be found in Hannaert et al. (2003).

In summary, cyt*b*, cox2 and 3 are needed in the insect stage, the ND subunits in one of the bloodstream stages and ATP6 probably in all lifecycle stages. Massive editing occurs in most of the transcripts needed. *T. brucei*, with its highly complex lifecycle, is the trypanosome species with the most abundant panediting of all organisms studied so far. Whether these two facts – complexity of lifecycle and high levels of panediting – are somehow interconnected will be discussed in the following.

The mass of kDNA encoding all these transcripts and their cognate gRNAs amounts to 10^{10} kDa (40% of the total cellular DNA), and is organized in a large network of thousands of catenated circles. Like all mt DNAs, kDNA is far less stable than chromosomal DNA. Mutations, large-scale deletions and even total loss of all kDNA, e.g. due to asymmetric divisions of heteroplasmic populations, have been observed frequently. This gave rise to the hypothesis that the kDNA network present in trypanosomatids evolved to counter these kinds of rapid mitotic segregations (Borst 1991). In some cases, so-called dyskinetoplastic strains (like *Trypanosoma equiperdum* ATCC 30023) can even survive in specific hosts. This poses a very challenging problem to parasites, because it is likely that individuals missing large parts of their kDNA (except for the relatively small part which is still needed) could easily out-compete normal individuals in the host (see Fig. 4A). In this case, the parasites would effectively die with their host. In the light of these considerations, it could be envisaged that increasing the amount of editing (i.e. splitting genes in a cryptogene containing the GAC sequence and multiple gRNA genes containing the information for the U sequence, which are then shuffled in such a way that parts encoding the complete message are found in many different locations) would be strongly selected for. Individual trypanosomes can no longer have large deletions in their kDNA because of the concomitant loss of essential gRNAs which have to be retained in the host (whether insect or mammal, in the case of *T. brucei*).

Fig. 4 Schematic representation of a panediting evolution model. Individual kinetoplastids are represented by single multicoloured circles of mt DNA with two or three genes only: a 'green' gene required for growth in the 'green' environment (*large green circle*; e.g. the mammal in the case of *T. brucei*), a 'red' gene required for growth in the 'red' environment (*large red circle*; e.g. the fly in the case of *T. brucei*) and a 'blue' gene required in both. The *arrows* underneath the environments indicate multiple replication rounds selecting the most rapidly multiplying individuals. Reduced generation times result from loss of mt DNA segments (note reduction in circle size) not under selective pressure in the respective environments. *Crosses* indicate absence of kinetoplastids still able to switch to the other environment. **A** Potential loss of host infectivity in parasites with mt genomes under limited selection. **B** Panediting: retention of host infectivity in parasites as a result of extensive gene fragmentation and scattering of gene fragments over the mt genome. Editing leads to gene fragmentation because it reduces the size of the 'gene' to the GAC sequence only, with the 'U sequence' encoded by multiple gRNAs placed in between other gene fragments. This in turn will make large-scale deletions result in loss of viability because the production of proteins which *are* under selective pressure will now also be compromised. Genome size is not affected by multiple replication rounds and, though small-scale mutations may still abound, some organisms will be able to pass the selective barrier (indicated by *). Adapted from Speijer (2006)

A. No Pan-editing

B. Pan-editing

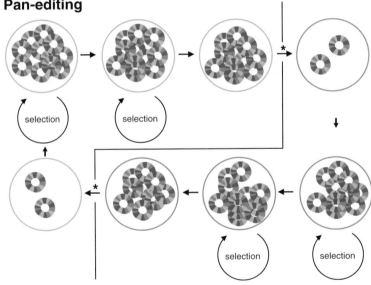

The essentials of this model are shown in Fig. 4A and B. To explain the central idea in the most general way: panediting evolved as a mechanism to minimize the danger of irreversible loss of mt genes encoding proteins necessary only during part of the kinetoplastid lifecycle. Loss of these genes from the mt DNA would be rampant because of the 'replication advantage' of deletion of large parts of the kinetoplast DNA during intraspecific competition in every 'new' environment making different demands on the mitochondrion. Editing is an effective way of making such large-scale deletions improbable, because (in order of relative importance):

1. It allows the intermixing of 'gene fragments' (multiple gRNAs and the cryptogene) in such a way that large-scale deletions will result in loss of gene fragments necessary for different lifecycles, and thus give rise to a non-viable organism,
2. It makes the presence of *all* the gRNAs (containing the information for the U sequence) necessary for the expression of such a 'gene', and
3. It *reduces* the local size of the original gene to that of the cryptogene (encoding the GAC sequence only).

Ad 1. The best candidates for gene fragmentation are the genes encoding proteins which are 'always' needed, rather than those needed only in some environments. The local presence of an essential 'part' of these genes will protect the mt DNA in that region against loss during strong intraspecific competition in *all* environments. That would mean that the mRNAs encoding a mitoribosomal protein such as RPS12 or, in the case of *T. brucei*, ATP6 should be the ones to be extensively edited. This is exactly what is observed (see Fig. 2). Indeed, RPS12 is the only gene found so far which encodes a transcript extensively edited in both kinetoplastid suborders, trypanosomatids and bodonids. Of course, also transcripts such as cox3 or ND7 could be in need of panediting as a result of the kind of evolutionary pressures envisaged in this model, because their fragmentation would protect the kDNA in those lifecycles in which they are essential (see above). In essence, the fragmentation of mt genes and the 'random' mixing of the resulting fragments can be seen as an extreme form of protective genetic linkage.

Ad 2. The different gRNAs all contain a part of the information needed for the production of the complete protein, making every single one of them indispensable. This is nicely illustrated by the fact that panediting has 3' to 5' directionality with the information of almost every single gRNA being crucial for the generation of the anchor sequence for the next, culminating in the final generation of an in-frame start codon (see Fig. 5D). Thus, every loss of a gRNA leads to the loss of a functional transcript, making gRNA genes perfectly suitable in rendering large-scale mt DNA deletions impossible.

Ad 3. A puzzling aspect of the kinetoplastid RNA editing process is the asymmetric distribution of U insertions and deletions. In *T. brucei*, we find a total of 2,965 instances of U insertions in the nine panedited transcripts, compared to only 318 U deletions, with insertion always far in the majority in every single panedited transcript. This is also found in the three transcripts with limited editing, having 64 insertions in total with only four deletions. The fact that insertions far outnumber deletions can be explained by the current model, as this distribution translates into a strong overall *reduction* of local 'gene' size in the case of cryptogenes, with a concomitant increase of small parts of the gene distributed over the mt genome in the form of gRNA genes.

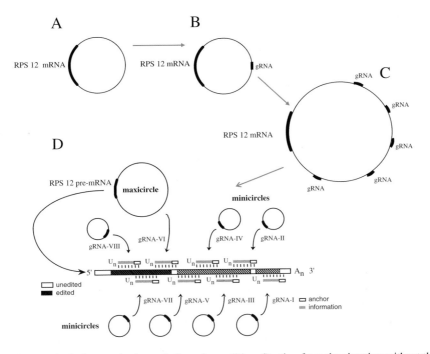

Fig. 5 Hypothetical stages in the evolution of panediting. Starting from the situation without the need for editing of the transcript (**A**), a single necessary gRNA is first encoded on the same DNA molecule (**B**). More gRNAs spread over the mt genome, possibly with a concomitant increase in size (**C**). Finally, minicircles encode most of the different gRNAs which are needed to end up with the fully edited mRNA (**D**). In **D**, the RPS12 mRNA transcribed from the maxicircle of *L. taren-tolae* is depicted to illustrate the current situation and the sequential action of gRNAs in panediting. *Red arrows* indicate possible developments during evolution. The three editing domains are *hatched*. The eight gRNAs involved in editing of the transcript and their genomic origin are indicated. Anchor regions are *white*, informational regions *grey*. Each of the three domains is edited separately with a 3′ to 5′ polarity. For further details, see the text (adapted and extended from Sloof and Benne 1997)

Whether the reduction in size (from a complete gene to its GAC 'skeleton') of genes which are not protected by selective pressure under certain conditions (such as cox3 in the *T. brucei* bloodstream form) significantly reduces their chance to be lost from the local population as a result of strong intraspecific competition remains a matter of debate. It should be stressed that these aspects (the requirement that *all* cognate gRNAs are present and the fact that insertions far outnumber deletions) do not make any sense in models where editing is invoked as an extra level of expression control or as a way of making more proteins from a 'single gene'.

Further observations seem to support the model. Extensive editing is indeed most common in kinetoplastid organisms which live under high evolutionary pressure, like members of the genus *Trypanosoma*. The role of gRNAs as 'checkpoints' for mt kDNA integrity, i.e. ensuring that large-scale deletions are not viable, could also be performed by evenly distributed tRNA genes (compare, e.g. their distribution in the circular human mt DNA; Attardi 1985). They are essential in every lifecycle, similarly to the two mt rRNAs, 9S and 12S, and the mt-encoded ribosomal protein RPS12. In *T. brucei*, *L. tarentolae* and *C. fasciculata*, however, tRNAs are not encoded by the kinetoplast DNA and are imported from the cytoplasm (Hancock and Hajduk 1990; Schneider et al. 1994). This could be the case for all kinetoplastid protozoa; to date, information regarding the genome location of mt tRNAs in the other flagellates is lacking. Mt tRNA import thus also seems to be a rather ancient trait, implying that tRNAs were not there to play a role in supplying 'checkpoints for kDNA integrity'. Could large-scale kDNA deletions really make such a difference in speed of multiplication of the individual organism? If we look, e.g. at the rapidly dividing, long slender form of *T. brucei*, only the parts encoding the ATP6 protein, the rRNAs of 9 and 12S, the RPS12 protein and the mt origins of replication seem to be under selective pressure. This makes up less than 30% of the coding capacity. A reduction of ~70% of the ~40% share of all DNA in the organism (kDNA is really huge, which is why it was the first extranuclear DNA discovered; Ziemann 1898) would mean ~30% reduction of the overall 'replication load'. Clearly, in a rapidly dividing population without panediting, trypanosomes retaining their capacity for switching between lifecycles would be at a serious disadvantage. Presumably, when panediting evolved, the kDNA, though large, was not that massive: paradoxically, editing, selected to counteract the temporary selective disadvantage of 'useless' kDNA, seems to have increased kDNA size.

Another interesting feature can be observed when we examine the known subunits of complex I. *T. brucei* capable of multiplying in the Tsetse fly can come only from the short stumpy population in the mammalian host. This population has an operational complex I for which at least six, and possibly eight mt-encoded subunits are necessary (ND3–5, ND7–9 and CR3 and 4), 4–6 of which are extensively edited (Feagin 2000; see also above). In terms of the model, this means that the fly will take up a population still able to synthesize a functional complex I, and therefore most likely possessing a full complement of genetic information. This in turn relies heavily on the integrity of the kDNA. Thus, at least some individuals will still be able to generate a completely functional respiratory chain starting with complex II in the insect. At first glance, the model would seem somewhat counterintuitive here: a gene such as cox3 is 'chopped into little pieces' to make it *less* likely of being lost during rapid growth under non-selective conditions. Many trypanosomes will indeed have lost the capability to synthesize a functional COX III protein during growth in the mammalian host but a sample taken up by the fly will have a much higher chance of still containing a few individuals which have retained this ability, in contrast to the situation without panediting in which they would have been outcompeted.

8.5 The Appearance of gRNAs

How did the gRNAs evolve? Several proposals have been made. The most simple model proposes that they originated by recruitment of random sequences (Simpson and Maslov 1999). Another model is based on kinetoplastid cox2 gRNA, the only gRNA sequence known so far which works (only) in cis because it is located in the 3' untranslated region of the cox2 mRNA. As this interaction needs only a very small anchor region and the transcription of the 'gRNA' is linked to its target, it has been proposed that all gRNAs started as intramolecular guides (Golden and Hajduk 2005). Yet other workers interpret them as the indirect product of mt DNA recombination, presumably coupled with the emergence of minicircles (Horton and Landweber 2002). As a result of recombination, a (small) circle encoding part of an mRNA with both an origin of replication and sequences allowing transcription is envisaged. It would be easier to choose between all these models if we knew the copy number of the mt DNA circles at the time gRNA genes started to evolve. The problem with the intramolecular gRNA model is the fact that the sequence 'comes out of nowhere', that there is only one example and that cox2 gRNA probably evolved when panediting already existed (as can be deduced from Fig. 2). The most parsimonious scenario is that as a result of recombinations, (parts of) complementary strands of certain protein coding genes had the possibility to mutate, giving rise to the forerunners of gRNA genes. Being part of 'normal' circles, replication and transcription of such sequences, possibly as parts of polycistronic messengers, would have been ensured.

8.6   *Discussion*

This model for the acquisition of panediting in kinetoplastids also helps to explain some other general features of kinetoplastids. The kinetoplastids have many biochemical aspects which are absolutely unique. They infect a wide range of hosts due to their rapid speciation and metabolic flexibility. As shown in Fig. 4, every time the kinetoplastids have to readjust to a different environment (indicated with an * in Fig. 4B), only a few of them in the sample still have all the genetic information necessary. However, this does not mean that they have not changed genetically. Evolutionarily speaking, we are looking at a founder effect with every switch between hosts. Again taking *T. brucei* as (the most extreme) example: from a rapidly dividing population of slender forms under strong selective pressure resulting from the sequential outgrowth of clonal populations with ever changing variant surface glycoproteins (VSGs; see Borst and Rudenko 1994; Vanhamme et al. 2001), a tiny sample is taken up by the fly and, from this sample, only a part will be able to thrive in the insect. This would ensure rapid speciation and the development of unexpected, 'weird' biochemical properties. This recurring founder effect could also help explain the acquisition of (limited) RNA editing in the first place: the 'genetic drift' phase of the Covello and Gray model clearly becomes less unlikely. Of course, mt functioning is integrated in the overall cellular metabolism. Its rapid evolution could accommodate large changes in kinetoplastid metabolism, such as the development of 'turbo-type' glycolysis and the unique glycosome in which it occurs (Hannaert et al. 2003). RNA editing has not been found in the diplonemids, the kinetoplastids sister group, but there has been a recent report of the cox1 gene in parasitic *D. papillatum* being split up in ~250-bp fragments: again, this points to the possibility that spreading genes throughout the mt DNA is essential for different kinds of parasites (Marande et al. 2005). Why do diplonemids and, for that matter, other parasites possibly confronted with the problem of rapid mt DNA loss not all end up with panediting? There seem to be several possibilities. First of all, only the kinetoplastids appear to have acquired limited RNA editing in the first place. Secondly, only organisms which really compete severely with each other in at least one of their lifecycle stages would need to fragment their genes in mt DNA. Thirdly, the kinetoplastid lineage could already have lost all its tRNA genes to the nucleus, making it more vulnerable to large-scale mt deletions. Editing and tRNA import from the nucleus in kinetoplastids could be linked, but we do not know which triggered which. Finally, it could be that kDNA was already disproportionally large at the time editing evolved, thus laying a larger claim on the organisms' resources than in parasitic organisms with a relatively small mt genome. Another question is related to the fact that *T. brucei* is something of an exception with its (even for kinetoplastids) highly elaborate lifecycle. Could the ancestors in which panediting has supposedly evolved really have encountered such differing demands regarding mt contributions to their overall metabolism? An important point: the discussion of the model has been illustrated with examples taken from the parasitic lifestyle of the kinetoplastids, but any free-living ancestor in a periodically changing environment in which the organism uses highly differential mt functioning (Cavalier-Smith 1997) would be subject to the same kind of selective pressures, favouring fragmentation of mt genes *as long as strong intraspecific competition would occur*.

It is not easy to critically evaluate this model for the evolution of panediting. The present-day situation for (most of) the kinetoplastids does not reflect the evolutionary pressures faced by their 'panediting ancestor'. This is strongly reflected by the fact that panediting is becoming ever less extensive in the kinetoplastid lineage. We can, however, reconstruct the most likely evolutionary history in terms of the proposed theory (see Fig. 5) and assess whether this is supported by extensive studies of other kinetoplastids. Starting from the situation with a single necessary gRNA probably encoded on the same DNA molecule, an increasing number of essential gRNAs would have spread over the mt genome, possibly with a concomitant increase in size. The large increase in coding capacity needed and the intricate regulation of transcription necessary could have been reasons for the emergence of minicircles encoding most of the different gRNAs needed to generate fully edited mRNAs. The presence of several thousand small minicircles of hundreds of different sequence classes, over and above the 50 maxicircles, could have made replication and equal distribution upon division so complex that this, in turn, lead to the evolution of the kDNA network to combat minicircle loss (Borst 1991). Both minicircles and networks thus presumably evolved later, in response to coding capacity demands of the panediting mechanism (Blom et al. 2000), reflecting its drawbacks. Although minicircles are thus seen as a later development, the intermingling of gRNAs on minicircles in the trypanosomatid order is again strongest in *T. brucei* (encoding 2–5 gRNA genes; Hong and Simpson 2003). The 180-kb gRNA encoding circles found in the early-diverging cryptobiid kinetoplastid,

T. borreli (Yasuhari and Simpson 1996), come closer to the hypothetical ancestral state illustrated in Fig. 5.

What other predictions does the model make? One important 'in vivo' prediction of the model involves the correlation between the complexity of the lifecycle of a kinetoplastid and the amount of panediting. Such a prediction is fraught with difficulties, however, because the present-day level will also reflect the complexity the organisms' ancestors encountered during their evolutionary history. Extensive studies of lifecycle complexity, evolutionary history and amount of mt panediting have to be combined to check the model in this respect. Did panediting really evolve as the unlikely keeper of the kinetoplast genome? Let's wait for what the lab has to tell us!

9   Concluding Remarks

Seeing the diversity in evolutionary histories and uses of RNA editing, both in organisms displaying as well as in mechanisms used in RNA editing, one could wonder whether there are also general patterns. First of all, nature loves complexity. Every additional level of complexity in an organism gives selection extra possibilities to shift into new and unexpected directions (in itself a form of 'evolvability'). Secondly, one should never underestimate the power of genetic drift. Crucial in this respect is the population size, small populations going into unexpected directions much more easily than our intuition would allow for. Within this context, the possible occurrence of RNA editing in bacteria could be discussed: maybe their larger effective population sizes would exclude this. Higher-order processes, such as sexual propagation (with its concomitant creation of a species genepool), could very well be the original legacy of small population sizes. Thirdly, panediting in kinetoplastids nicely illustrates that evolution is 'trying to hit a moving target' (and that the 'trying' is part of the 'moving'). Last but not least, panediting in kinetoplastids also reflects 'conflicting' tendencies in evolutionary innovations: it speeds up evolution but, at the same time, makes gene loss much less likely due to the extreme genetic linkage it entails. Again, parallels with sex are obvious: this opens up new possibilities by rapidly forming myriads of combinations from the genepool but is also a conservative force because 'any … enterprising new evolutionary direction is held in check by the swamping effect of sexual mixing' (see Dawkins 2004). RNA editing thus illustrates, in its microcosmos, the great capability of very simple mechanisms to generate all the unpredictable wonders of molecular biology. To paraphrase Charles Darwin: 'there is a wonder in this view of RNA editing, with its several powers, having been originally started with few forms; and from so simple a beginning endless forms most beautiful and most wonderful have been, and are being evolved' (Darwin 1859).

Acknowledgements I thank Bob Meek, Jeffrey Ringrose, Paul Sloof and, especially, Rob Benne for critical reading of the manuscript.

References

Arts GJ, Benne R (1996) Mechanism and evolution of RNA editing in Kinetoplastida. Biochim Biophys Acta 1307:39–54

Attardi G (1985) Animal mitochondrial DNA: an extreme example of genetic economy. Int Rev Cytol 93:93–145

Baldauf SL (2003) The deep roots of eukaryotes. Science 300:1703–1706

Bass BL (2002) RNA editing by adenosine deaminases that act on RNA. Annu Rev Biochem 71:817–846

Benne R (ed) (1993) RNA editing, the alteration of protein coding sequences of RNA. Ellis Horwood, Chichester, UK

Benne R, van den Burg J, Brakenhoff JPJ, Sloof P, van Boom JH, Tromp MC (1986) Major transcript of the frameshifted coxII gene from trypanosome mitochondria contains four nucleotides that are not encoded in the DNA. Cell 46:819–826

Bienen EJ, Saric M, Pollakis G, Grady RW, Clarkson AB Jr (1991) Mitochondrial development in *Trypanosoma brucei brucei* transitional bloodstream forms. Mol Biochem Parasitol 45:185–192

Bishop KN, Holmes RK, Sheehy AM, Malim MH (2004) APOBEC-mediated editing of viral RNA. Science 305:645

Blom D, de Haan A, van den Berg M, Sloof P, Jirku M, Lukeš J, Benne R (1998) RNA editing in the free-living bodonid *Bodo saltans*. Nucleic Acids Res 26:1205–1213

Blom D, de Haan A, van den Burg J, van den Berg M, Sloof P, Jirku M, Lukeš J, Benne R (2000) Mitochondrial minicircles in the free-living bodonid *Bodo saltans* contain two gRNA gene cassettes and are not found in large networks. RNA 6:121–135

Blum B, Bakalara N, Simpson L (1990) A model for RNA editing in kinetoplast mitochondria: 'guide' RNA molecules transcribed from maxicircle DNA provide the edited information. Cell 60:189–198

Börner GV, Pääbo S (1996) Evolutionary fixation of RNA editing. Nature 383:225

Borst P (1991) Why kinetoplast DNA networks? Trends Genet 7:139–141

Borst P, Rudenko G (1994) Antigenic variation in African trypanosomes. Science 264:1872–1873

Byrne EM, Gott JM (2004) Unexpectedly complex editing patterns at dinucleotide insertion sites in *Physarum* mitochondria. Mol Cell Biol 18:7821–7828

Cavalier-Smith T (1997) Cell and genome coevolution: facultative anaerobiosis, glycosomes and kinetoplastan RNA editing. Trends Genet 13:6–9

Covello PS, Gray MW (1993) On the evolution of RNA editing. Trends Genet 9:265–268

Cruz-Reyes J, Zhelonkina AG, Huang CE, Sollner-Webb B (2002) Distinct functions of two RNA ligases in active *Trypanosoma brucei* RNA editing complexes. Mol Cell Biol 22:4652–4660

Darwin C (1859) On the origin of species by means of natural selection. Reprint Penguin, London, UK

Davidson NO (1993) Apolipoprotein B mRNA editing: a key controlling element targeting fats to proper tissue. Ann Med 25:539–543

Dawkins R (2004) The ancestor's tale. Weidenfeld & Nicolson, London, UK

Feagin JE (2000) Mitochondrial genome diversity in parasites. Int J Parasitol 30:371–390

Feagin JE, Abraham JM, Stuart K (1988) Extensive editing of the cytochrome *c* oxidase III transcript in *Trypanosoma brucei*. Cell 53:413–422

Fernandes AP, Nelson K, Beverley SM (1993) Evolution of nuclear ribosomal RNAs in kinetoplastid protozoa: perspectives on the age and origins of parasitism. Proc Natl Acad Sci USA 90:11608–11612

Gerber AP, Keller W (2001) RNA editing by base deamination: more enzymes, more targets, new mysteries. Trends Biochem 26:376–384

Golden DE, Hajduk SL (2005) The 3′-untranslated region of cytochrome oxidase II mRNA functions in RNA editing of African trypanosomes exclusively as a cis guide RNA. RNA 11:29–37

Gott JM, Visomirski LM, Hunter JL (1993) Substitutional and insertional RNA editing of the cytochrome *c* oxidase subunit 1 messenger RNA of *Physarum polycephalum*. J Biol Chem 268:25483–25486

Grivell LA (1993) Plant mitochondria 1993 – a personal overview. In: Brennicke A, Kück U (eds) Plant mitochondria. VCH Verlagsgesellschaft, Weinheim, pp 1–14

Hancock K, Hajduk SL (1990) The mitochondrial tRNAs of *T. brucei* are nuclear encoded. J Biol Chem 265:19208–19215

Hannaert V, Bringaud F, Opperdoes FR, Michels PAM (2003) Evolution of energy metabolism and its compartmentation in Kinetoplastida. Kinetopl Biol Dis 2:1–30

Hanson MR (1996) Protein products of incompletely edited transcripts are detected in plant mitochondria. Plant Cell 8:1–3

Hausmann S, Garcin D, Delenda C, Kolakofsky D (1999) The versatility of paramyxovirus RNA polymerase stuttering. J Virol 73:5568–5576

Hirose T, Sugiura M (2001) Involvement of a site-specific trans-acting factor and a common RNA-binding protein in the editing of chloroplast mRNAs: development of a chloroplast in vitro RNA editing system. EMBO J 20:1144–1152

Hoekstra RF (2005) Why sex is good. Nature 434:571–572

Hong M, Simpson L (2003) Genomic organization of *Trypanosoma brucei* kinetoplast DNA minicircles. Protist 154:265–279

Horton TL, Landweber LF (2000) Evolution of four types of RNA editing in myxomycetes. RNA 6:1339–1346

Horton TL, Landweber LF (2002) Rewriting the information in DNA: RNA editing in kinetoplastids and myxomycetes. Curr Opin Microbiol 5:620–626

Kapushoc ST, Alfonso JD, Rubio MA, Simpson L (2000) End processing precedes mitochondrial importation and editing of tRNAs in *Leishmania tarentolae*. J Biol Chem 275:37907–37914

Koslowsky DJ, Bhat GJ, Perrolaz AL, Feagin JE, Stuart K (1990) The MURF3 gene of *T. brucei* contains multiple domains of extensive editing and is homologous to a subunit of NADH dehydrogenase. Cell 62:901–911

Kung SS, Chen YC, Lin WH, Chen CC, Chow WY (2001) Q/R RNA editing of the AMPA receptor subunit 2 (GRIA2) transcript evolves no later than the appearance of cartilaginous fishes. FEBS Lett 509:277–281

Laforest MJ, Roewer I, Lang BF (1997) Mitochondrial tRNAs in the lower fungus *Spizellomyces punctatus*: tRNA editing and UAG 'stop' codons recognized as leucine. Nucleic Acids Res 25:626–632

Laforest MJ, Bullerwell CE, Forget L, Lang BF (2004) Origin, evolution, and mechanism of 5′ tRNA editing in chytridiomycete fungi. RNA 10:1191–1199

Landweber LF (1992) The evolution of RNA editing in kinetoplastid protozoa. BioSystems 28:41–45

Landweber LF, Gilbert W (1993) RNA editing as a source of genetic variation. Nature 363:179–182

Landweber LF, Gilbert W (1994) Phylogenetic analysis of RNA editing: a primitive genetic phenomenon. Proc Natl Acad Sci USA 91:918–921

Lavrov DV, Brown WM, Boore JL (2000) A novel type of RNA editing occurs in the mitochondrial tRNAs of the centipede *Lithobius forficatus*. Proc Natl Acad Sci USA 97:13738–13742

Leigh J, Lang BF (2004) Mitochondrial 3′ tRNA editing in the jakobid *Seculamonas ecuauadoriensis*: a novel mechanism and implications for tRNA processing. RNA 10:615–621

Levanon EY, Hallegger M, Kinar Y, Shemesh R, Djinovic-Carugo K, Rechevi G, Jantsch MF, Eisenberg E (2005) Evolutionarily conserved human targets of adenosine to inosine RNA editing. Nucleic Acids Res 33(4):1162–1168

Lin S, Zhang H, Spencer DF, Norman JE, Gray MW (2002) Widespread and extensive editing of mitochondrial mRNAs in dinoflagellates. J Mol Biol 320:727–739

Lonergan KM, Gray MW (1993) Editing of transfer RNAs in *Acanthamoeba castellanii* mitochondria. Science 259:812–816

Lu B, Hanson MR (1994) A single homogenous form of ATP6 protein accumulates in petunia mitochondria despite the presence of differentially edited *atp6* transcripts. Plant Cell 6:1955–1968

Lu B, Hanson MR (1996) Fully edited and partially edited *nad9* transcripts differ in size and both are associated with polysomes in potato mitochondria. Nucleic Acids Res 24:1369–1374

Lukeš J, Arts GJ, Van den Burg J, de Haan A, Opperdoes F, Sloof P, Benne R (1994) Novel pattern of editing regions in mitochondrial transcripts of the cryptobiid *Trypanoplasma borreli*. EMBO J 13:5086–5098

Lynch M, Koskella B, Schaack S (2006) Mutation pressure and the evolution of organelle genomic architecture. Science 311:1727–1730

Madison-Antenucci S, Grams J, Hajduk SL (2002) Editing machines: the complexities of trypanosome RNA editing. Cell 108:435–438

Marande W, Lukeš J, Burger G (2005) Unique mitochondrial genome structure in diplonemids, the sister group of kinetoplastids. Eukaryot Cell 4:1137–1146

Maslov DA, Simpson L (1994) RNA editing and mitochondrial genomic organisation in the cryptobiid kinetoplastid protozoan *Trypanoplasma borreli*. Mol Cell Biol 14:8174–8182

Maslov DA, Sturm NR, Niner BM, Gruszynski ES, Peris M, Simpson L (1992) An intergenic G-rich region in *Leishmania tarentolae* kinetoplast maxicircle DNA is a pan-edited cryptogene encoding ribosomal protein S12. Mol Cell Biol 12:56–67

Miller D, Mahendran R, Spottswood M, Ling M, Wang S, Yang N, Costandy H (1993) RNA editing in mitochondria of *Physarum polycephalum*. In: Benne R (ed) RNA editing, the alteration of protein coding sequences of RNA. Ellis Horwood, Chichester, UK, pp 87–103

Nolan DP, Voorheis HP (1992) The mitochondrion in bloodstream forms of *Trypanosoma brucei* is energized by the electrogenic pumping of protons catalysed by the F_1F_0-ATPase. Eur J Biochem 209:207–216

Ojala D, Montoya J, Attardi G (1981) tRNA punctuation model of RNA processing in human mitochondria. Nature 290:470–474

Palladino MJ, Keegan LP, O'Connell MA, Reenan RA (2000) A-to-I pre-mRNA editing in *Drosophila* is primarily involved in adult nervous system function and integrity. Cell 102:437–449

Polson AG, Bass BL, Casey JL (1996) RNA editing of hepatitis delta virus antigenome by dsRNA-adenosine deaminase. Nature 380:454–456

Raoult D, Audic S, Robert C, Abergel C, Renesto P, Ogata H, La Scola B, Suzan M, Claverie J-M (2004) The 1.2-megabase genome sequence of Mimivirus. Science 306:1344–1350

Schnaufer A, Ernst NL, Palazzo SS, O'Rear J, Salavati R, Stuart K (2003) Separate insertion and deletion subcomplexes of the *Trypanosoma brucei* RNA editing complex. Mol Cell 12:307–319

Schneider A, Martin J, Agabian N (1994) A nuclear encoded tRNA of *Trypanosoma brucei* is imported into mitochondria. Mol Cell Biol 14:2317–2322

Scott J (1995) A place in the world for RNA editing. Cell 81:833–836

Seeburg PH (1996) The role of RNA editing in controlling glutamate receptor channel properties. J Neurochem 66:1–5

Shaw JM, Feagin JE, Stuart K, Simpson L (1988) Editing of kinetoplastid mitochondrial mRNAs by uridine addition and deletion generates conserved amino acid sequences and AUG initiation codons. Cell 52:401–411

Shields DC, Wolfe KH (1997) Accelerated evolution of sites undergoing mRNA editing in plant mitochondria and chloroplasts. Mol Biol Evol 14:344–349

Shikanai T (2006) RNA editing in plant organelles: machinery, physiological function and evolution. Cell Mol Life Sci 63:698–708

Simpson L, Maslov DA (1999) Evolution of the U-insertion/deletion RNA editing in mitochondria of kinetoplastid protozoa. Ann NY Acad Sci 870:190–205

Simpson L, Thiemann OH, Savill NJ, Alfonzo JD, Maslov DA (2000) Evolution of RNA editing in trypanosome mitochondria. Proc Natl Acad Sci USA 97:6986–6993

Simpson AGB, Lukeš J, Roger AJ (2002) The evolutionary history of kinetoplastids and their kinetoplasts. Mol Biol Evol 19:2071–2083

Skuse GR, Cappione AJ, Sowden M, Metheny LJ, Smith HC (1996) The neurofibromatosis type I messenger RNA undergoes base-modification RNA editing. Nucleic Acids Res 24:478–486

Sloof P, Benne R (1997) RNA editing in kinetoplastid parasites: what they do with U. Trends Microbiol 5:189–195

Smith HC, Snowden MP (1996) Base-modification mRNA editing through deamination – the good, the bad and the unregulated. Trends Genet 12:418–424

Sommer B, Köhler M, Sprengel R, Seeburg PH (1991) RNA editing in brain controls a determinant of ion flow in glutamate-gated channels. Cell 67:11–19

Speijer D (2006) Is kinetoplastid pan-editing the result of an evolutionary balancing act? IUBMB Life 58:91–96

Thiemann OH, Maslov DA, Simpson L (1994) Disruption of RNA editing in *Leishmania tarentolae* by the loss of minicircle-encoded guide RNA genes. EMBO J 13:5689–5700

Tomita K, Ueda T, Watanabe K (1996) RNA editing in the acceptor stem of squid mitochondrial tRNA^{Tyr}. Nucleic Acids Res 24:4987–4991

Tsudzuki T, Wakasugi T, Sugiura M (2001) Comparative analysis of RNA editing sites in higher plant chloroplasts. J Mol Evol 53:327–332

Turelli P, Trono D (2005) Editing at the crossroad of innate and adaptive immunity. Science 307:1061–1065

van der Spek H, Speijer D, Arts GJ, van den Burg J, van Steeg H, Sloof P, Benne R (1990) RNA editing in transcripts of the mitochondrial genes of the insect trypanosome *Crithidia fasciculata*. EMBO J 9:257–262

van der Spek H, Arts GJ, Zwaal RR, van den Burg J, Sloof P, Benne R (1991) Conserved genes encode guide RNAs in mitochondria of *Crithidia fasciculata*. EMBO J 10:1217–1224

Vanhamme L, Pays E, McCulloch R, Barry JD (2001) An update on antigenic variation in African trypanosomes. Trends Parasitol 17:338–343

Vidal S, Curran J, Kolakofsky D (1990) A stuttering model for paramyxovirus P mRNA editing. EMBO J 9:2017–2022

Visomirski-Robic LM, Gott JM (1995) Accurate and efficient insertional RNA editing in isolated *Physarum* mitochondria. RNA 1:681–691

Visomirski-Robic LM, Gott JM (1997) Insertional editing in isolated *Physarum* mitochondria is linked to RNA synthesis. RNA 3:821–837

Volchkov VE, Becker S, Volchkova VA, Ternovoj VA, Kotov AN, Netesov SV, Klenk HD (1995) GP mRNA of Ebola virus is edited by the Ebola virus polymerase and by T7 and vaccinia virus polymerases. Virology 214:421–430

Yasuhari S, Simpson L (1996) Guide RNAs and guide RNA genes in the cryptobiid kinetoplastid protozoan, *Trypanoplasma borreli*. RNA 2:1153–1160

Yokobori S, Pääbo S (1995a) Transfer RNA editing in land snail mitochondria. Proc Natl Acad Sci USA 92:10432–10435

Yokobori S, Pääbo S (1995b) tRNA editing in metazoans. Nature 377:490

Yokobori S, Pääbo S (1997) Polyadenylation creates the discriminator nucleotide of chicken mitochondrial tRNA(Tyr). J Mol Biol 265:95–99

Ziemann H (1898) Eine Methode der Doppelfärbung bei Flagellaten, Pilzen, Spirillen und Bakterien, sowie bei einigen Amöben. Zentralbl Bakteriol Parasitenkd Infektionskr Hyg 24:945–955

Index

Printing: Krips bv, Meppel
Binding: Stürtz, Würzburg